NUTRITION AND DIET RESEARCH PROGRESS

Additional books in this series can be found on Nova's website under the Series tab.

Additional E-books in this series can be found on Nova's website under the E-books tab.

NUTRITION AND DIET RESEARCH PROGRESS

SATURATED FATS

METABOLISM, DISEASE RISKS AND PUBLIC AWARENESS

NUTRITION AND DIET RESEARCH PROGRESS

SATURATED FATS

METABOLISM, DISEASE RISKS AND PUBLIC AWARENESS

JOSEPH P. LANGELLA
EDITOR

Nova Science Publishers, Inc.
New York

For permission to use material from this book please contact us:
Telephone 631-231-7269; Fax 631-231-8175
Web Site: http://www.novapublishers.com

Additional color graphics may be available in the e-book version of this book.

Library of Congress Cataloging-in-Publication Data

Saturated fats : metabolism, disease risks, and public awareness / editor,
Joseph P. Langella.
p. ; cm.
Includes bibliographical references and index.
ISBN 978-1-61209-566-0 (hardcover)
1. Fat--Metabolism. 2. Fat--Physiological effect. 3. Lipids in
nutrition. I. Langella, Joseph P.
[DNLM: 1. Dietary Fats--metabolism. 2. Dietary Fats--adverse effects.
QU 86]
QP752.F3S28 2011
612.3'97--dc23

2011024371

Published by Nova Science Publishers, Inc. † New York

Contents

Preface

This new book presents current research in the study of the metabolism, disease risks and public awareness of saturated fats. Topics discussed include the epidemiological evidence and dietary intervention studies regarding the relationship of fats, carbohydrates and risk of coronary heart disease; the modulation of dietary conjugated linoleic acid (CLA) effects by saturated fats; effects of short and long-term saturated fat-enriched diet on the ceramide and neutral lipids accumulation in the insulin responsive tissues of rats and fatty acids and immune system functions.

Chapter 1 – The physiological actions of various macronutrients are the subject of continued address because of their possible role in the cause of coronary heart disease and other chronic disorders such as diabetes mellitus. Diet influences serum lipoproteins and total cholesterol concentration has been known for years. Numerous dietary strategies have been proposed that claimed to be best for the reducing the coronary heart disease risk. Initially, the national dietary recommendation was mainly focused on a reduction in fat intake. In common man's mind, the word "dietary fat" was embedded as synonymous with obesity and heart disease, where as the word "low fat" and "fat free" were taken as synonymous with heart health. This lead to a huge switch over to the consumption of fat free or low fat foods rich in high amounts of refined carbohydrates and sugar. However, paradoxically, such dietary recommendations over the years resulted in the increased prevalence of unintended health consequences like obesity, insulin resistance, metabolic syndrome and type 2 diabetes mellitus, which are known to increase the risk for coronary heart disease. Thus, indicating that the replacement of fats with carbohydrates to reduce heart disease was based on little scientific evidence and much less attention was given to the expected degree of risk reduction. It is now also increasingly appreciated that different fats have different health effects and total fat intake does not increase cardiovascular disease risk. For instance, results from many epidemiological and controlled clinical trials have indicated that replacing saturated fat with unsaturated fats is more effective in reducing the cardiovascular disease risk than simply reducing total fat consumption. In this chapter, the authors discuss the epidemiological evidence and dietary intervention controlled studies regarding the relationship of fats, carbohydrates and risk of coronary heart disease with particular emphasis on, " Is a low saturated fat, low carbohydrate diet a strategy to reduce coronary heart disease risk"?

Chapter 2 – Today's unhealthy diet and lifestyles promote excessive intake of saturated fats. High fat diets, possessing inappropriate portions of unsaturated *versus* saturated fatty acids (SFA), may directly increase the prevalence of obesity worldwide. In contrast to SFA, dietary polyunsaturated fatty acids (PUFA) have been recognized as capable of reducing the accumulation of abdominal adipose tissue and lipids in the liver and serum, and also to alleviate hypertension and type 2 diabetes, through the regulation of many enzymes involved in both lipid and glucose metabolisms. One unsaturated fatty acid that has attracted much of the scientific community interest is the conjugated linoleic acid (CLA), a group of minor fatty acids of ruminant origin with potential beneficial effects for long term human health. Interestingly, some studies revealed that CLA is more beneficial in controlling adiposity and blood lipids when supplemented to a diet rich in saturated fat *versus* unsaturated fat. In fact, a complex interaction between dietary manipulations (such as, type and levels of fat) and the body fat-reducing action of CLA has been already reported. However, research performed so far has not adequately compared the biological effects of CLA and the respective molecular mechanisms of action when supplemented to diets containing distinct sources of saturated fat.

In this book chapter, the authors report their own scientific journey concerning the disclosure of CLA biological effects and the underlying mechanisms of action upon dependence on saturated fat dietary matrices with differential lipid compositions. Two distinct experimental trials were carried out with normal-weight Wistar and obese Zucker rats fed saturated fat based diets supplemented with CLA isomers (*cis*(*c*)9,*trans*(t)11 and *t*10,*c*12, which are the known physiologically active isomers), separately or combined. New (molecular) insights towards CLA action were exploited focusing, in particular, on lipid and glucose metabolisms, inflammatory responses, apoptosis and involved signaling pathways, as well as biomembrane properties associated with cellular trafficking events.

Chapter 3 – High-fat diet is a significant risk factor for the development of insulin resistance and Alzheimer`s disease (AD). The link between dietary saturated fat-induced insulin resistance and neurodegeneration and AD has been determined. However, the precise mechanisms underlying these dietary influences are not completely understood. The authors investigated effects of dietary saturated fats and palmitic acid on ceramide and neutral lipids accumulation in the insulin responsive tissues and insulin-induced glucose uptake. 2-week feeding of rats with saturated fat increased sphingomyelin (SM) and decreased phosphatidylcholine levels and did not change the ceramide and neutral lipids and free fatty acids contents in the liver. The short-term diet also increased the palmitic and oleic acids and decreased the arachidonic acid contents. The dietary saturated fat, chronically added to the food, was found to increase free fatty acids as well as fatty acid metabolites (ceramide and neutral lipids) contents in such insulin key target tissues as liver, muscle and brain. Similar results were obtained in the experiments with palmitate- or C2-ceramide-treated cell and tissue cultures. Treatment of the isolated liver hepatocytes, diaphragm strips and cortex slices of adult rats which had a free access to a standard chow diet with palmitic acid increased the ceramide and neutral lipids contents in the cells and tissues and decreased the insulin-induced glucose uptake by the cells. The palmitic acid effect on insulin-stimulated glucose transport in the tissues studied could be mimicked by the addition of short chain (C2) ceramide to the culture medium. Taken together, their results suggest that long-term exposure of adult rats to the dietary saturated fat leads to ceramide accumulation in the tissues inducing ceramide synthesis from the palmitate and activating the SMases activities. It is rather the ceramide accumulation in the cells than neutral lipids or fatty acids accumulation is responsible and sufficient to inhibit the insulin-induced glucose uptake by the cells.

Chapter 4 – A major transport form of lipids in insect haemolymph are diacyglycerols (DGs), predominantly bound to carrier protein lipophorins, in contrast to vertebrates where the lipid transport is established by free fatty acids (FAs). Mobilization of the DGs from the fat body and their transport to the target effectors, mostly to working muscle cells, is under the control of adipokinetic hormones (AKHs). The AKHs play a crucial role in the control of insect energy metabolism - they behave as typical stress hormones by stimulating catabolic reactions, making energy more available, while inhibiting synthetic reactions. The DG molecular species and their FA composition were investigated by electrospray mass spectrometry (ESI-MS) in haemolymph of *Locusta migratoria* and *Pyrrhocoris apterus* after the application of their innate AKHs (Locmi-AKH-I, AKH-II and AKH-III in *L. migratoria*, and Pyrap-AKH-I and Peram-CAH-II in *P. apterus*). Mobilization of DGs from the fat body after injection of AKH was not uniform. AKHs mobilized the DGs selectively with preference of those possessing the C18 and C16 FAs, especially 18:1 in *L. migratoria* and 18:2 in *P. apterus.* The fat body FAs with a carbon chain longer than 18 did not participate in the mobilization. In *L. migratoria,* Locmi-AKH-I preferred release of DGs with unsaturated FAs, while AKH-II and AKH-III stimulated DG release with saturated FAs. In *P. apterus,* significant quantitative differences between Pyrap-AKH-I and Peram-CAH-II were also shown: the former preferred linoleic acid (18:2), the later oleic acid (18:1). Heterologous assays, where the AKHs from one species were applied to another one, suggested that the activation of DGs and/or FAs was rather species- than hormone-specific. The relationship between the AKH action and FA distribution in DGs are summarized and discussed in detail.

Chapter 5 – Composition of human tissues lipids and their metabolism are of paramount importance for the growth, development and long-term human health. Lipids are the major source of energy in the body and supply essential lipid-soluble vitamins and polyunsaturated fatty acids (PUFA) that are required in relatively high amounts during growth and life. Lipids affect the composition of membrane structures and modulate membrane functions as well as the functional development of the central nervous system. Some long-chain PUFA serve as precursors for bioactive lipid mediators.

The principal long-chain polyunsaturated fatty acids, arachidonic acid (20:4n-6, AA), eicosapentaenoic acid (20:5n-3, EPA), and docosahexaenoic acid (C22:6n-3, DHA), i.e. products of chain elongation and desaturation of the essential fatty acids, linoleic acid (18:2n-6, LA) and α-linolenic acid (18:3n-3, ALA), are important for growth and maintaining various biological functions in the human organism. Long chain PUFA of the n-3 series are essential for early development of the central nervous system, whereas at later stages of life, they are beneficial for cardiovascular health, reducing platelet aggregation and triacylglycerol (TAG) levels; they may play a role in the prevention of coronary artery disease, hypertension and arthritis, as well as some inflammatory and autoimmune disorders and some cancers. EPA, dihomo-γ-linolenic acid (20:3n-6) AA, and DHA are precursors of prostaglandins, thromboxanes and leukotrienes, which are powerful regulators of numerous cell functions, playing positive role in the pathophysiology of thrombosis and atherosclerosis, as well as immunologic processes and inflammation. The percentage of EPA + DHA in red blood cells, also known as the Omega-3 Index, has recently been proposed as a risk marker for death from coronary heart disease.

Chapter 6 – The potential role of certain fatty acids on immune system functions has been intensively investigated during the last thirty years. Fatty acids are characterized by exerting various biological functions on the cells from mammalian. In addition to their crucial role as main source of energy or as structural substances, different types of fatty acids have been considered as modulating agents of immune system in both animals and humans. Based on this preliminary argument, different mechanisms of action have been proposed in order to explain these events: i) the capacity of fatty acids to alter the membrane fluidity because they are incorporated in the plasma membrane phospholipids, ii) the formation of lipid peroxides that may be toxic to the cells, iii) the changes in the production of lipid mediators such as eicosanoids or iv) the direct intervention in gene expression. Recently, numerous studies have determined their contribution as inducers or inhibitors of apoptosis. Therefore, this last event may constitute an important explanation to clarify the reasons whereby certain fatty acids modulate the immune system. Thus, the most immunosuppressive fatty acids are derived from long-chain *n*-3 polyunsaturated fatty acids (mainly contained in fish oil) and from *n*-9 monounsaturated fatty acids (mainly contained in olive oil). These fatty acids are involved in the reduction of lymphocyte proliferation, cytokine production, natural killer (NK) cell activity, and alteration of surface molecules. The modification of these immune functions leads to a suppression of innate and adaptive immunity. Hence, these actions acquire a significant importance in the application of these fatty acids in the diminution of inflammatory disorders, which are characterized by an overactivation of immune system, or in the reduction of tumor development. In spite of the relevance of several fatty acids of the diet as beneficial agents in the human health, an excess of determined types of fatty acids may be involved in an increase of the host susceptibility to infectious microorganisms. Taking into account the consequences derived from the administration of dietary lipids, here the authors will review the role of dietary lipids as biological agents capable of modulating the immune system function. Moreover, the authors will study the mechanisms of action and the immune parameters susceptible to be modified after the incorporation of certain fatty acids in the diet. A second part of this series will analyze the clinical consequences derived from the administration of certain dietary lipids, which will be based on the intervention of fatty acids in immunonutrition; their application in clinical nutrition, their functions as modifier agents of immune resistance or their role in cancer aetiology.

In: Saturated Fats
Editor: Joseph P. Langella

ISBN: 978-1-61209-566-0

Chapter I

Dietary Macronutrients, Plasma Lipoproteins and Cardiovascular Disease Risk

***Manmeet Singh*[*1,3], *Kota J. Reddy*[1], *Joey R. Bangit*[1], *Richard R. Batsell*[2] *and Kamal Kotak*[3]**

[1]Reddy Cardiac Wellness, Sugar Land, TX
[2]Rice University, Houston, TX
[3]UCSF Fresno, Department of Internal Medicine, Fresno, CA

Abstract

The physiological actions of various macronutrients are the subject of continued address because of their possible role in the cause of coronary heart disease and other chronic disorders such as diabetes mellitus. Diet influences serum lipoproteins and total cholesterol concentration has been known for years. Numerous dietary strategies have been proposed that claimed to be best for the reducing the coronary heart disease risk. Initially, the national dietary recommendation was mainly focused on a reduction in fat intake. In common man's mind, the word "dietary fat" was embedded as synonymous with obesity and heart disease, where as the word "low fat" and "fat free" were taken as synonymous with heart health. This lead to a huge

* Corresponding Author: Manmeet Singh M.D - 155 N Fresno Street, Fresno, CA 93711

switch over to the consumption of fat free or low fat foods rich in high amounts of refined carbohydrates and sugar. However, paradoxically, such dietary recommendations over the years resulted in the increased prevalence of unintended health consequences like obesity, insulin resistance, metabolic syndrome and type 2 diabetes mellitus, which are known to increase the risk for coronary heart disease. Thus, indicating that the replacement of fats with carbohydrates to reduce heart disease was based on little scientific evidence and much less attention was given to the expected degree of risk reduction. It is now also increasingly appreciated that different fats have different health effects and total fat intake does not increase cardiovascular disease risk. For instance, results from many epidemiological and controlled clinical trials have indicated that replacing saturated fat with unsaturated fats is more effective in reducing the cardiovascular disease risk than simply reducing total fat consumption. In this chapter, we discuss the epidemiological evidence and dietary intervention controlled studies regarding the relationship of fats, carbohydrates and risk of coronary heart disease with particular emphasis on, " Is a low saturated fat, low carbohydrate diet a strategy to reduce coronary heart disease risk"?

Introduction

Substantial data is available to indicate that diet influences lipoprotein and total cholesterol levels. In recent years, however, attention has shifted to individual lipoprotein and their sub- fractions. Fats, cholesterol and carbohydrates are three dietary components that have been identified to alter total cholesterol, lipoproteins and their sub-fractions levels. Several lines of evidence indicate that the types of fat have a more important role in determining the risk of cardiovascular disease than the total amount of fat. Saturated fats are major lipoprotein raising fats that have been shown to increase the risk of coronary heart disease (CHD) [1]. Conversely, the role of dietary cholesterol in the causation of the mass hypercholestremia and raising lipoproteins is not universally accepted because of reports that human's response to dietary cholesterol varies and only a fixed proportion of cholesterol is absorbed from the human intenstines [2]. Considering that carbohydrates have neutral effects on total cholesterol and lipoprotein levels in the earlier studies, consumption of fat free or low fat food rich in high amounts of refined carbohydrates and sugar has grown dramatically in recent years [3-4]. However, numerous studies evaluated that replacing fats with carbohydrates have suggested that there is a little, if any, benefit to be expected. In this chapter, we will discuss on whether a low saturated fat, low carbohydrate diet could be the primary dietary intervention for reducing CHD risk while reviewing the epidemiological evidence and dietary intervention studies regarding the relationship of fats, carbohydrates and CHD risk.

Dietary Cholesterol, Serum Cholesterol and Cardiovascular Disease

The fact that dietary cholesterol is highly atherogenic in many species has encouraged the view that high cholesterol consumption may possibly be atherogenic in humans [5-7]. The proposed hypothesis that a high intake of cholesterol increases the number of atherogenic LDL particles is based on the mechanism that increasing the cholesterol content of cells suppresses the LDL receptor activity and down regulates the synthesis of LDL receptor. This change in LDL receptor activity results in an increase of circulating levels of atherogenic LDL particles. However, this has not been consistently demonstrated in various studies done in humans. For instance, Ginsberg et al [8] found no influence of dietary cholesterol on LDL metabolism in humans who were essentially non responders to excessive dietary cholesterol. On the other hand, another study done by Packard et al [9] reported that raising the cholesterol intake causes both an increase in the production rate of LDL and a decrease in the fractional catabolic rate for LDL. The fall in the fractional catabolic rate is suggestive of a decrease in LDL-receptor activity, where as the increase in the production rate of LDL can be attributed to the increase hepatic synthesis of Apo-B containing lipoprotein such as very low density lipoproteins leading to the formation of LDL particles. On the other hand, the decrease in the LDL receptor could also allow the rapid conversion of VLDL to LDL in a same manner as in the patients of primary hypercholestremia who are devoid of LDL receptors resulting in the increase production of LDL particles. Thus, it appears that it is not necessary to induce overproduction of hepatic Apo B lipoproteins to increase input of LDL.

The quantitative correlation between dietary cholesterol intake and serum cholesterol levels seems to be minor and remains a matter of some dispute. Now it is increasingly appreciated that effects of dietary cholesterol on raising the atherogenic lipoproteins are less pronounced in humans than previously thought. In fact, a large number of observational studies have suggested that dietary cholesterol has little or no effect on serum cholesterol levels [10-12]. Among these were the Honolulu Heart Study, the Puerto Rico Heart-Health Program, the Lipid Research Clinics Prevalence Follow-up Study, the Nurses Health Study, the Alpha Tocopherol Beta-Carotene Cancer Prevention Study, the Hawaii

Cardiovascular Study, and the Framingham Study [13-23]. In an argument to these studies, many researchers believe that marked variability in responsiveness to dietary cholesterol exists and this could be a possible factor determining serum cholesterol levels. Katan and colleagues in 1986 reported consistent differences in cholesterol serum levels to dietary cholesterol of healthy individuals [24]. In a followup study, the same group reported that such intervariability to dietary cholesterol highly correlated to the saturated fat content in the diet [25]. Such interindividual intervariability could also be partly attributed to the differences in the cholesterol absorption. In one study that measured cholesterol, absorption varied widely in range from 26-80% with a mean absorption of 56%. The same study reported that cholesterol absorption is significantly higher in the African American population than all racial groups [26]. Similar rates of absorption were found in other studies that hinted that such inter-variability seemingly exists in humans. However, interestingly, a study reported the rate of cholesterol absorption decreases with increasing intake of dietary cholesterol. In this study, 18 normal subjects were fed with two test meals containing different amounts of cholesterol. The study reported that the efficiency of cholesterol absorption declines appreciably with modest increases in the cholesterol doses [27]. This study draws attention to the fact that the effect of dietary cholesterol on serum cholesterol is a complex issue, and the present hypothesis of high dietary cholesterol resulting in high serum cholesterol needs more appropriate evidence.

In respect to the association of dietary cholesterol and cardiovascular disease risk, the evidence is not ultimate. A significant positive association among dietary cholesterol and cardiovascular disease was found in some epidemiologic studies, but not in others. The Nurses Health Study[18] followup found a weak and nonsignificant positive association between dietary cholesterol and risk for cardiovascular disease (relative risk for each increase of 200mg /1000kcal= 1.12, 95 % confidence interval 0.91-1.40). The Western Electric Study [28-29], which prospectively followed 1824 middle aged men for a period of 25 years found the association between dietary cholesterol content and CHD. In the Zutphen study [30], dietary cholesterol intake per 1000 kcal was positively related to coronary heart disease. However, in the same study, energy intake was inversely related to CHD risk. They found that the relationship between dietary cholesterol and cardiovascular disease was not significant when energy intake was taken into consideration. Similarly, findings from EURODIAB IDDM study [31] and Ascherio et al [32] illustrated that the association of dietary cholesterol with CHD was no longer significant after the adjustment of confounding factors. The results from Seven Countries Study related 25 years of cardiovascular disease death rates with consumption of saturated fatty acid($r>0.80$), trans fatty acid($r=0.78$) and dietary cholesterol [33]. However, findings were not able to determine the independent effects of saturated fatty acid and dietary cholesterol on serum cholesterol and cardiovascular mortality.

Perhaps the strongest evidence that the impact of dietary cholesterol on CHD in general is negligible comes from the investigations done on eggs. The Framingham study group investigated the association of CHD and eggs that provide high amount of cholesterol without a high amount of total fat and saturated fats [34]. The study estimated the egg consumption on 912 subjects and found that there was no relationship whatsoever between egg consumption and cardiovascular heart disease. Similarly, in an extensive analysis of egg consumption and incidence of CHD among the 117,933 apparently healthy subjects in the Nurses Health Study follow-up, revealed no association between egg consumption and risk of CHD in either men or women [35]. In another case, a control study of the frequency of egg consumption was also not significantly associated with the risk of CHD in women. Thus, the findings from egg studies and inferences from other observational studies provided convincing evidence that the effect of dietary cholesterol alone on the development of cardiovascular disease is minor and is rather promoted by certain eating patterns(high saturated fat, high trans fat) related to high cholesterol intake. Nevertheless, despite years of research, the topic of dietary cholesterol and its association with CHD risk is still very controversial and is far from reaching a consensus.

Types of Dietary Fats

Dietary Fats are composed of hydrocarbon chains of fatty acids with a terminal methyl and carboxyl group. There are less than twenty fatty acids that are quantitatively important in human diet. Out of these twenty, two (palmitic acid and oleic acid) often account for around 65 % of total fatty acid intake. The most commonly occurring fatty acids are grouped as saturated fats, monounsaturated fats and polyunsaturated fats, depending upon the presence and number of double bonds. On the basis of the evidence provided in early dietary reports by Keys [3] and Hegsted [4], dietary fats have been long stereotyped as leading risk factors for CHD; and are considered to be a number one forbidden food in the American health. As a result, subtle shifts developed over the years in dietary recommendations that led to an emphasis on the reduction in fats intake in general, with less distinction about the type of fat. Following these recommendations, although the mortality associated with cardiovascular disease declined dramatically over the years, the degree to which these dietary changes contributed to decline in CHD mortality still remains unsettled. It is now increasingly recognized that different types of fats have different health effects and reducing total fat per se does not attribute to increased cardiovascular disease risk. This is confirmed in the meta analysis results of numerous controlled feeding studies of the effects of different dietary fatty acids on serum cholesterol levels and CHD risk. More recently, using 14 year follow-up data from nurses health study Hu and colleagues [18] conducted a prospective analysis of dietary fats and cardiovascular disease among 80,082 women aged 34-59. The findings suggested that estimated replacement of five percent of energy from saturated fat by unsaturated fats would reduce risk by 42 percent (23-56, $P < 0.001$), and the replacement of two percent of energy from trans fat by unhydrogenated unsaturated fats would reduce the risk by 53 percent (34-67, $P<0.001$). They also reported that total fat intake was not significantly related to cardiovascular disease risk (for five percent energy 1.02, 0.97-1.07, $p=0.55$). Consequently, these findings suggest that replacing saturated and trans fats with unsaturated fats is more useful in reducing cardiovascular disease risk than slashing overall fat intake.

Saturated Fats, Plasma Lipids and Mechanisms of Action

All 3 classes of fatty acids (saturated, monounsaturated and polyunsaturated) elevate high density lipoprotein cholesterol (HDL-C) relative to other macronutrients (especially carbohydrates). However, this proportional increase of HDL-C by saturated fats is abolished by its adverse effects on low density lipoprotein cholesterol (LDL-C). Keys and Hegsted [3-4] were among the first investigators who established that saturated acids increase total cholesterol concentrations and LDL-C; and this fact has not been contradicted by subsequent investigations (Fig 1). Saturated fats raise the LDL-C more than anything else. The mechanisms by which saturated fatty acids raise LDL-C level remains to be clarified. The most widely accepted theory is that saturated fat increases LDL-C by decreasing receptor mediated catabolism which adversely affect the LDL levels and have little effect on production rate.

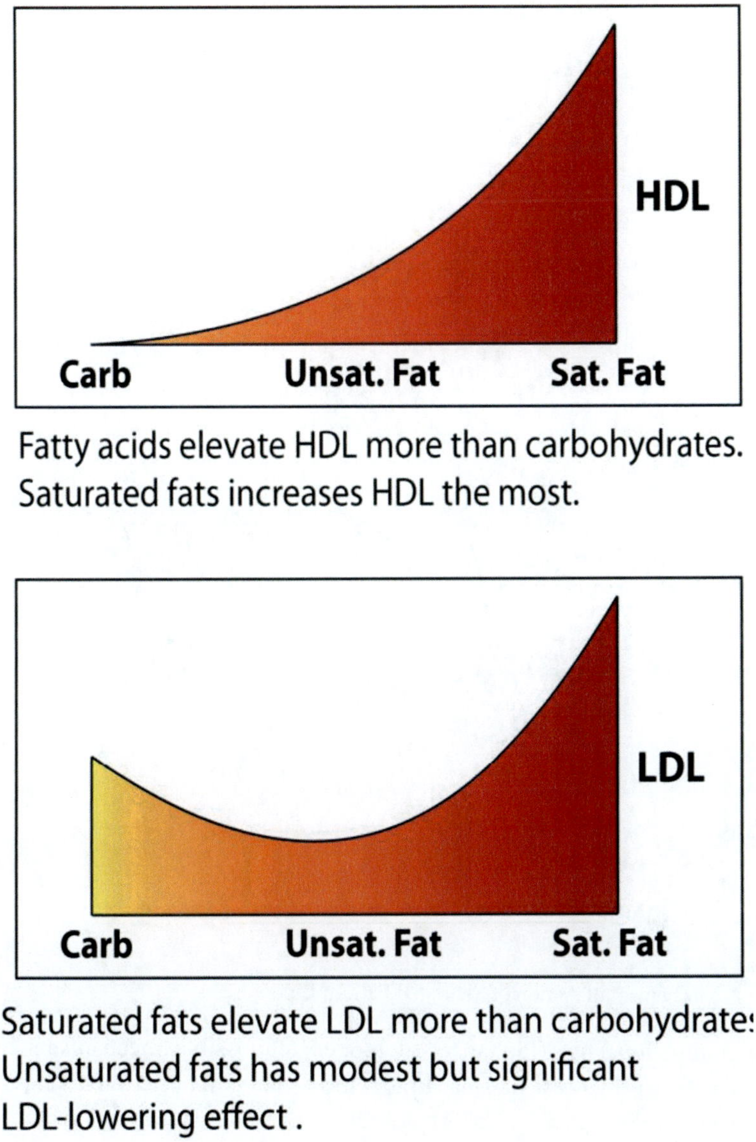

Figure 1. Dietary Macronutrients & Effect on Lipoproteins.

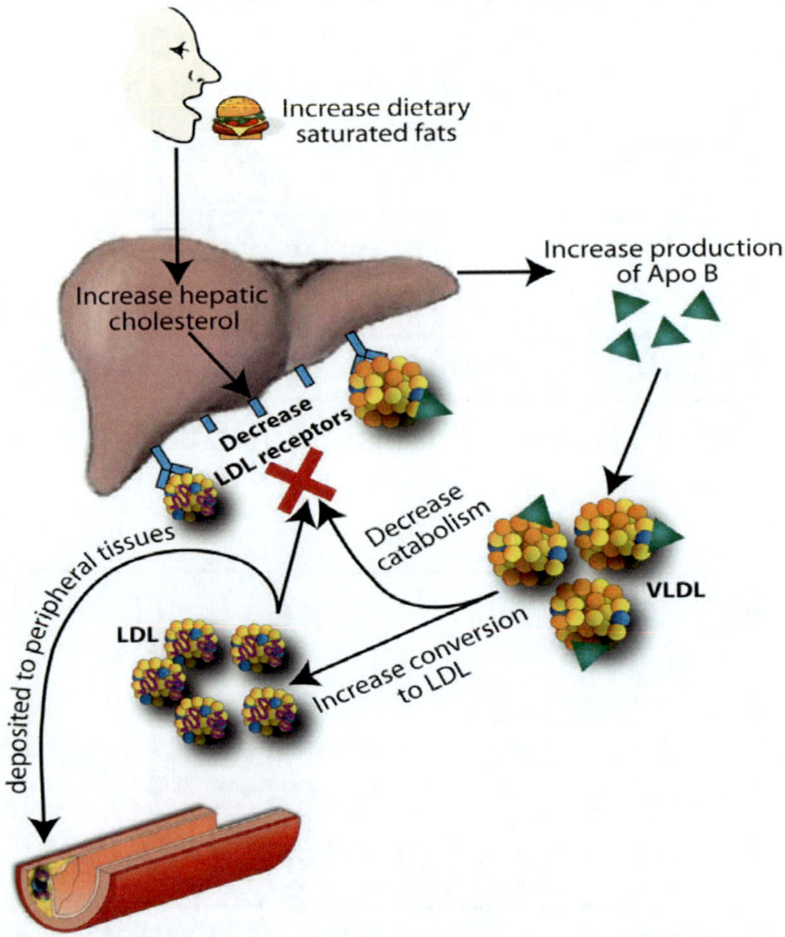

Figure 2. Mechanism of Action of Saturated Fats on Lipoproteins.

A study done by Matthan et al looked at possible mechanisms by which saturated fat influences plasma lipids [36]. They reported that elevations in circulating levels of cholesterol are not due to a rise in cholesterol synthesis, but is most likely secondary to an impairment of the catabolic pathway of lipoproteins. Conversely, research in many animal and human studies does not show consistent evidence that the only mechanism for increasing LDL-C by saturated fats is reduced clearance through a LDL-receptor pathway [37]. Alternatively, another possible mechanism could be the enhanced synthesis of ApoB containing lipoproteins resulting in the increase production of very low density lipoproteins (VLDL), eventually increasing LDL cholesterol [38] (Figure2). This mechanism, however, was also contradicted by the Sorci-Thomas et al study who reported that a diet high in saturated fatty acid does not increase the expression of mRNA for ApoB in monkeys [39]. Similarly, Shady and Dietschy reported that saturated fatty acid does not enhance the synthesis of ApoB containing lipoproteins. In the same study, researchers found no difference in ApoB containing lipoproteins after pre-feeding with high saturated fats [40]. The proof about the mechanism is far from fully revealed and gets further complicated by the fact that all saturated fatty acids do not affect plasma lipids and lipoproteins in the same manner. Hegsted et al and Grande et al proposed that stearic acid; a type of saturated fatty acid for example, has little effect on plasma cholesterol and LDL-C concentrations [41]. Grundy and Denke proposed this could be a result of rapid conversion of stearic acid in the body to oleic acid [42]. Furthermore, available evidence also suggests that the related effect of saturated fat on serum lipid and lipoproteins is dependent on the chain length. The longer the chain length, the smaller is the effect on the lipoproteins. Saturated fats with 12-16 carbon atoms tend to increase plasma and LDL-C levels more as compared to 18 carbon atoms saturated fats. The minimal effect in terms of raising LDL-C and decreasing HDL-C of the shorter chain fatty acids is attributed to their being absorbed directly into portal circulation, and that of 18:0 is attributed to its high rate of conversion to 18:1, a monounsaturated fatty acid. These findings, therefore, suggest that the effect of each type of saturated fat should be considered separately.

Major Types of Saturated Fatty Acid

Palmitic Acid

Palmitic acid is a 16 carbon atom saturated acid, which is believed to be a principal source of saturated fatty acid in the diet. It is a major saturated fatty acid in animal fats, but also occurs in vegetable oils. Palmitic acid is the first fatty acid produced during the lipogenesis and is the source from which the longer fatty acids can be produced. Keys and Hegsted reported that the rise in total cholesterol and LDL-C parallels the intake of palmitic acid [3-4]. Subsequent multiple studies also have clearly shown that palmitic acid increases both total and LDL cholesterol proportionally. There is convincing evidence provided by multiple studies, that the dietary intake of palmitic acid increases the risk of CHD. Watts et al investigated the dietary effects on the progression of coronary artery disease in men, and reported a significantly high correlation among higher intakes of palmitic acid and decrease in absolute width of coronary artery assessed by angiography [43]. Similarly, another study examined the relationship of serum fatty acid and the risk of coronary heart disease in a nested case-control study of 94 men with incident CHD and 94 men without incident CHD who were enrolled in the Usual Care group of the Multiple Risk Factor Intervention Trial between December 1973 and February 1976. In all multivariate models, levels of the palmitic acid (16:0) were directly associated with CHD risk (standardized odds ratio = 1.68; 95% confidence interval 1.10-2.55 in the model that adjusted for total plasma cholesterol level) [44].

Myristric Acid

Myristic acid is a 14 carbon atom saturated acid that is consumed in relatively small quantities, usually in the form of butter fat. Myristic acid is named after the nutmeg (Myristica fragrans). Nutmeg butter is 75% trimyristin, the triglyceride of myristic acid. Less data is available on the effects of myristic acid on plasma lipids and lipoproteins. In the early dietary studies done by Keys and Hegsted [3-4], a consensus was reached on the effects of myristic acid on total cholesterol, but Hegsted reported that myristic acid raises the total cholesterol even more than palmitic acid. The latter claim was bolstered in the Seven Countries Study that evaluated the association between the intake of individual fatty acids and dietary cholesterol in relation to serum cholesterol and 25-year mortality from coronary heart disease. For the individuals on saturated fatty acids, the average population intake of myristic acid along with lauric acid was most strongly related to the average serum cholesterol level ($r > 0.8$, $P < 0.001$). Not surprisingly, the study observed a strong positive association between 25-year death rates from coronary heart disease and that average intake of the four major saturated fatty acids, lauric, myristic, palmitic, and stearic acid ($r > 0.8$, $P < 0.001$); the trans fatty acid elaidic acid ($r = 0.78$, $P < 0.001$) [33]. Another epidemiological study that also studied the effect of individual saturated fat on CHD in the Costorican population reported a higher odd ratio of 1.51(95% confidence interval) for myristic and lauric acid [45]. Although myristic acids in studies were strongly associated with increased risk of MI, it should be noticed that they are consumed in small amounts and most of the saturated fat (87%) came from palmitic and stearic acids, is which derived mainly from red meat and fried foods.

Stearic Acid

Stearic acid is an 18 carbon atom saturated fat that is found in large quantities in cocoa butter. Similar to palmitic acid, stearic acid also occurs in animal and vegetable fats and is one of the major sources of saturated fats in the body. Unlike other saturated fatty acids, studies done in animals and humans strongly suggest that stearic acid does not raise the serum cholesterol. Even in their dietary report, Keys et al recommended that stearic acid should be deleted from a dietary mix of saturated fatty acids when calculating the serum response [3]. Likewise, the work of Ahrens et al found that cocoa butter (high in stearic acid) is less hypercholesteremic than regular butter and suggested stearic acid should not be included with saturated fats [46]. The possible proposed reasons for this effect must be considered. One possibility is that stearic acid is poorly absorbed and does not enter the body. Stearic acid is more insoluble than other fatty acids and could easily escape the absorption. Alternatively, the work of Grundy and Bonanome reported that stearic acid is very well absorbed in humans [47]. They proposed that the failure of stearic acid to raise serum cholesterol is most likely based on its rapid conversion of stearic acid to monounsaturated fatty acid (oleic acid).

Although these earlier studies raise uncertainties that stearic acid increases the serum cholesterol level as does other saturated acid, this lack of effect is not widely accepted simply because stearic acid is a saturated fatty acid. Moreover, available evidence linked stearic acid with risk of CHD. In a recent analysis of Nurses health study, Hu and colleagues found that dietary intakes of longer-chain saturated fatty acids (12:0-18:0) were each separately associated with a small increase in risk [18]. After an adjustment for coronary risk factors, the RR for a 1% energy increase from stearic acid was 1.19(1.02-1.37). Similarly, the 25 year mortality data from coronary heart disease in the Seven countries study [33] shows a strong positive association with all four saturated fatty acids($r > .80$, $p=0.001$). However, it could not independently determine RR for stearic acid. Additionally, stearic acid lowers HDL-C compared to monounsaturated or polyunsaturated fatty acids and HDL-C lowering effect was particularly strong among women. Aro et al reported that compared to myristic and palmitic acids, stearic acid reduced LDL-C concentration, but also reduced HDL-C concentration [48]. Thus, overall ratios of LDL to HDL remain the same. Stearic acid also impairs fibrinolysis, may activate factor VII and increases Lp(a) concentration. Thus, multiplicity of stearic acid effects in the body along with findings from the Nurses Health Study follow up (largest study of fat intake with four repeated assessments) proposes that a distinction between stearic acid and other saturated fats does not appear to be important in dietary advice to reduce CHD risk. Moreover, in typical diets, saturated fats and stearic acid are highly correlated.

Saturated Fats and Heart Disease Risk

Available literature showed a strong positive correlation between saturated fat intake and rates of CHD. This data, however, were seriously confounded by other aspects of diet, lifestyle factors, and economic development. Despite years of interest in diet-heart disease hypothesis, there are only a few prospective studies that examined the link between the saturated fat and CHD risk. Only a few of these prospective studies found a positive association among saturated fat intake and the risk of cardiovascular disease, as earlier studies of fat and CHD were limited by small sample size , inadequate dietary assessment, inadequate adjustment for total energy intake and types of fat or trans fat isomer intakes. The largest detail study of fat intake, with four repeated measurements, was carried out in a cohort group of 80,082 women of the Nurses Health Study which reported a weak positive association between saturated fat intake and the risk of CHD. The findings suggested that estimated replacement of five percent of energy from saturated fat by unsaturated fats would reduce risk by 42 percent (23-56, P < 0.001) [18]. In addition to this, few dietary trials with CHD end points have been conducted and most were among the patients with CHD. In these studies, two approaches were tested; one replacing the saturated fats with polyunsaturated fats with no change in total fat intake; the other lowered the total fat. Out of the trials involving the reduction of saturated fats with an increase in polyunsaturated fats, only in the Minnesota Coronary Survey, cardiovascular events were not significantly reduced despite a decrease in total serum cholesterol [49]. In the Los Angeles Veteran hospital study [50], the CHD rate was found to be reduced by 31% after 8 years, whereas in the Finish Mental Hospital study [51], the reduction reported in CHD was 43% over 6 years. However, the reduction in the Finnish study was probably related to the reduction in trans fat intake as soft margarine was substituted for stick margarine. On the other hand, the two secondary prevention trials carried out in non-institutionalized subjects reported contrasting results. In the Oslo Diet Heart Study, which included 21% energy replacement with polyunsaturated fat, found a significant reduction in serum cholesterol and major coronary events (fatal or non fatal MI) at the five year follow-up. At 11 years, fatal MI was reduced by 44% and total cardiovascular mortality by 14% [52-53]. Another trial done by Morris et al using high amount of soybean oil showed no effect on recurrent coronary events (non significant reduction of 12%) after 4 years, despite a 16% reduction in serum cholesterol at 6 months [54]. The trials testing a low total fat intake approach did not find a significant reduction in serum cholesterol or CHD events. In the MRC trial by Ball et al, among 123 male MI patients, 22% energy reduction of total fat showed no significant reduction in serum cholesterol or CHD events [55]. Similarly, the low-fat intervention group in the Diet and Reinfarction Trial (DART) reported no reduction in serum cholesterol or CHD events [56]. These findings from prospective studies and intervention trials in general demonstrate that replacing saturated fats with polyunsaturated fats is advantageous. However, it is not possible to determine which nutrient was responsible for the observed reduction in the CHD. Was it the reduction of saturated fat intake or was it the increase in total polyunsaturated fats intake? Was it the reduction in total serum cholesterol in these studies that resulted in observed reduction in CHD? Nevertheless, these trials support the clinical utility of overall dietary strategy of replacing saturated fatty acids with polyunsaturated fats in the heart disease prevention. Moreover, it supports the types of fatty acid consumed is more important then total fat reduction .

Saturated Fat, Vascular Inflammation and Endothelium

The development of atherosclerotic lesions is a complex process involving the endothelium, lipid rich macrophages (foam cells) and smooth muscle cells. Diets rich in fatty acids may directly influence many pathological events in the artery that promote the formation of atherosclerotic lesions. A key target for modulation is endothelium. Endothelial dysfunction is a critical event in atherogenesis that is implicated both in early and advanced atherosclerosis. It is characterized by a decreased bioavailability of nitric oxide and increased expression of pro-inflammatory cytokines and cellular adhesion molecules. Available work suggests that the type of fatty acid is an important factor that affects endothelial dysfunction. Nicholls et al, who investigated the influence of dietary fatty acids on vascular function and anti-inflammatory properties of HDL, demonstrated that an intake of a meal high in saturated fat results in an increased expression of ICAM-1 and VCAM-1 adhesion molecules secondary to the impairment of HDL anti-inflammatory properties [57]. Flow mediated dilation (FMD) also decreased 3 hours after the saturated meal by 2.2 ± 0.9 %($p<0.05$ compared with baseline) vs.

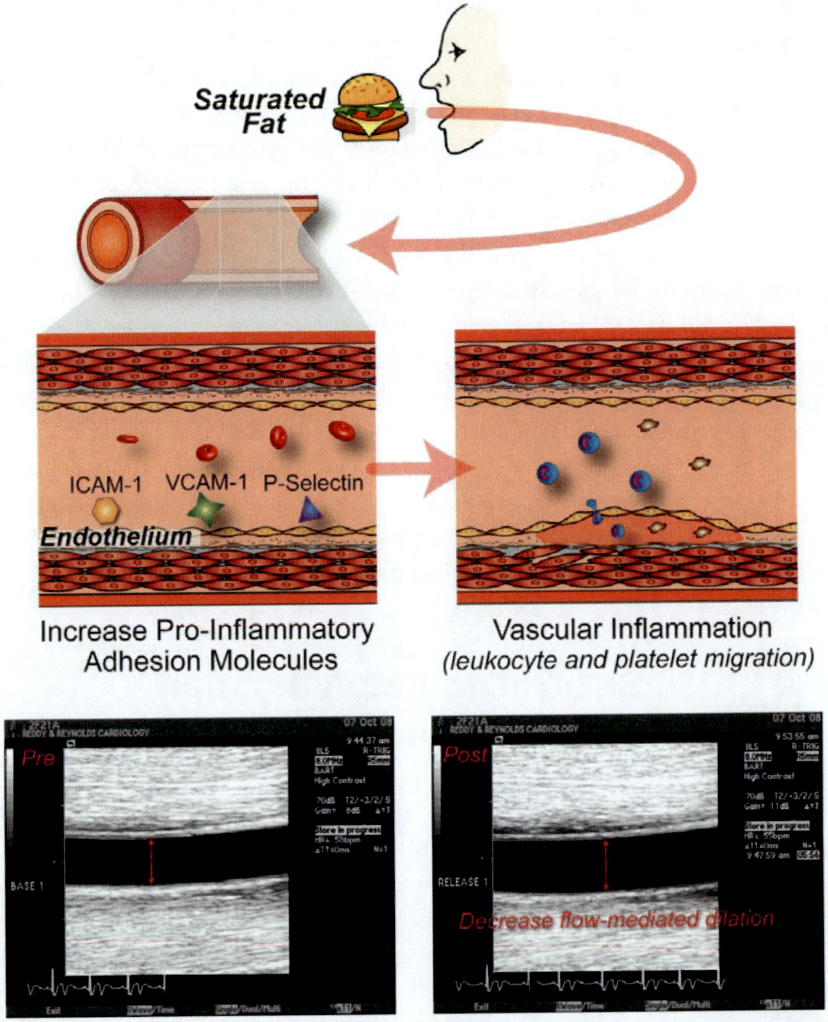

Figure 3. Saturated Fats and their Effect on Endothelial Function determined by Endothelium Dependent flow mediated Dilation

0.9±1% after the polyunsaturated meal. Similar effects of saturated fatty acids on post-prandial impairment of FMD (vascular reactivity) have been reported in some, but not all, studies [58-60]. Similarly, variable correlation has been reported between longer-term consumption of dietary fats and endothelial function in subjects with or without vascular risk factors. In another recent study,the adoption of a saturated fat-enriched diet for a 3 week period was associated with an impairment of FMD and elevation of P selectin in healthy young adults [61].

Vogel et al, conversely, reported that the consumption of a meal high in monounsaturated but not polyunsaturated fats was associated with a reduction in FMD during the post-prandial period in healthy volunteers [62]. The reason for such discrepancy is uncertain but may be attributed in a part to the fact that earlier studies were designed so that baseline was done in early morning and post-prandial testing was done 3-4 hours later. The physiologic improvement in FMD during the morning counteracted any changes due to fat challenge. Secondly, the protocol used assess the vascular function in earlier studies was not fully standardized. Thus, even though the mechanisms of underlying impaired endothelial function in the post-prandial states are multifactorial, the available literature raises the possibility that the intake of saturated fatty acids may contribute to the increase expression of pro-inflammatory adhesion molecules and endothelial dysfunction (Figure 3).

Saturated Fats and Insulin Resistance

Insulin resistance is the pathogenic link underlying different metabolic abnormalities clustering in metabolic syndrome [63-64]. Many environmental factors including dietary habits can induce insulin resistance. Consumption of high energy fatty foods has been long positively associated with insulin resistance. There is also enough evidence that relates the quality of fatty acids with insulin resistance independent of other factors. The impact of fatty acids consumption on insulin may be mediated through modifications in fatty acid composition of cell membrane [65-67]. A specific type of fatty acid profile in cell membrane might affect the action of insulin through several potential mechanisms including changing ion permeability, cell signaling and altering insulin receptor affinity. Higher saturated fatty acids content of the cell membrane increases insulin resistance. Thus, suggesting that diets rich in saturated fats may impair the insulin action on the cell membrane and rendering them insulin resistant. The other mechanism by which saturated fatty acid could alter insulin sensitivity partly depends on the activities of enzymes responsible for synthesizing, desaturating and elongating fat acids in the body as well as partly on the dietary fatty acid composition of the diet [68]. Indeed, the results from two of these reports suggest that insulin resistance is related to fatty acid pattern characterized by high proportions of palmitic acid (16:0), palmitoleic acid and low proportion of linoleic acid. Similar fatty acid pattern of high palmitic acid (saturated fatty acid) and low linoleic acid (unsaturated fatty acids) have been linked to insulin resistance in both prospective and cross- sectional studies [69]. Furthermore, various saturated fatty acids have different effects on insulin secretion in experiments and clinical studies. Long chain saturated fatty acids, in particular, have significant adverse effects on insulin sensitivity, resulting in insulin resistance (Figure 4).

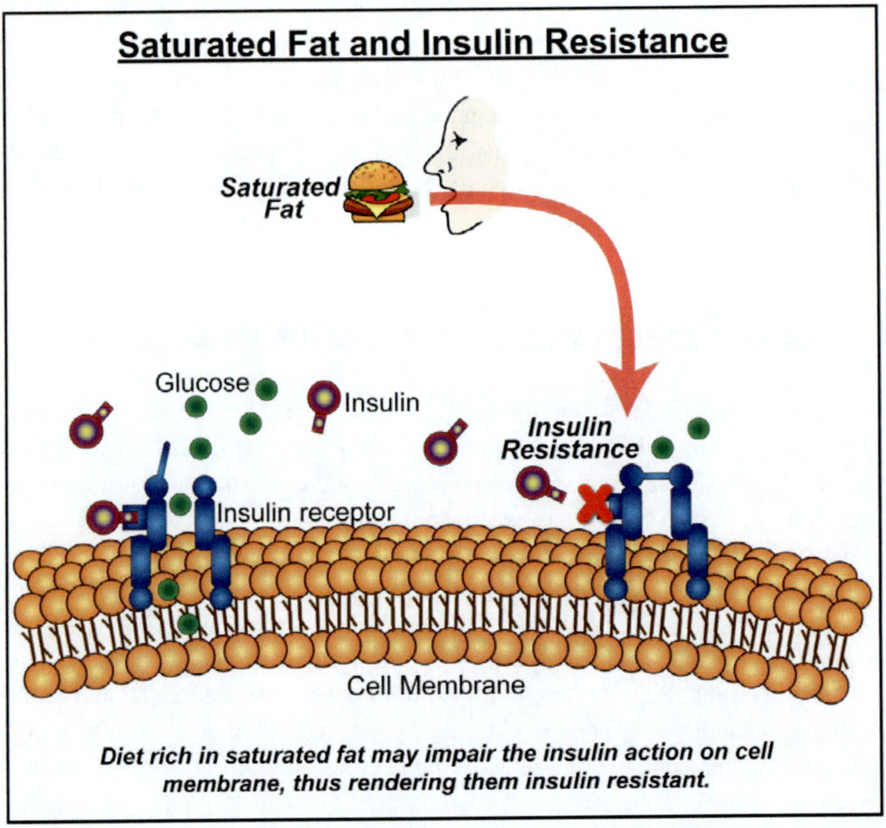

Figure 4. Saturated Fat & Insulin Resistance.

Epidemiological evidence and intervention studies clearly show that, in humans, saturated fat significantly worsens insulin-resistance, while monounsaturated and polyunsaturated fatty acids improve it through modifications in the composition of cell membranes which reflect, at least in part, dietary fat composition. A recent multicenter study (KANWU) has shown that shifting from a diet rich in saturated fatty acids (SFA) to one rich in monounsaturated fat improves insulin sensitivity in healthy people while a moderate alpha-3 fatty acids supplementation does not affect insulin sensitivity [70]. KANWU, the largest study done so far done included 162 healthy individuals from five different countries. During the SFA diet, insulin sensitivity decreased by 10%, whereas no change (+2%) occurred during the MUFA diet. Interestingly, this effect of SFA lowering insulin sensitivity was not seen in individuals with higher total fat intake (>37 % energy). The later finding from KANWU was in accordance with the results from another life style controlled study in which changes in estimated desaturase activities were related to the changes in insulin sensitivity only in those on diets below 35.5% energy consumption. Thus, indicating that fat quality and quantity both have a role to play in insulin sensitivity and the amount of quality fat should not be neglected in view of improving insulin sensitivity. Similarly, in another controlled, short-term (28 days) study in 59 healthy individuals, 29% lower insulin sensitivity after saturated fats versus monounsaturated fats was reported [71]. Additional evidence that insulin sensitivity is influenced by changing the fat quality was demonstrated in another randomized, controlled crossover study [72]. The study reported that a N-6 polyunsaturated fatty acid rich diet improved insulin sensitivity when substituted for a saturated –rich diet after only 5 weeks. Thus, suggesting the overall harmful role of saturated fats on insulin sensitivity over unsaturated fats.

Unsaturated Fats

On the basis of a number of double bonds, unsaturated fats are divided into monounsaturated fatty acids (MUFA) and polyunsaturated fatty acids (PUFA). An unsaturated fat particle is monounsaturated if it contains only one double bond where as it is poly unsaturated if it contains more than one double bond. Higher intake of unsaturated fats is correlated with lower rates for heart disease. The biological effects of MUFA depends whether the MUFA are in cis or trans configuration. Cis-MUFA has been considered neutral in relation to their effects on LDL and HDL cholesterol whereas trans-MUFA adversely effects plasma lipoproteins. Dietary PUFA, is also sub-classified as n-6 or n-3, depending on the location of the carbon atom involved in first double bond from the methyl end of chain.

Monounsaturated Fats and CHD Risk

That MUFA intake has an inverse association with CHD Risk has been reported in ecological studies. In particular, the mortality rate from CHD is very low in the traditional Mediterranean population that uses olive oil (monounsaturated fats) as the primary source of fat. Prospective cohort studies examining the MUFA intake and CHD risk are sparse. A study done by Pietinen et al [73] reported an inverse association after an adjustment of fat intake. On the other hand, two studies done by Posner et al & Esrey et al independently, have found an increased risk of CHD with higher intake of MUFA in younger but not in older participants [14,22]. However, neither of these two studies adjusted for the intake of other fat types. The Nurses Health Study found an inverse association between MUFA intake and CHD risk [18]. In the metabolic study exchanging carbohydrates for monounsaturated fats resulted in the favorable changes in plasma lipoproteins. Colette et al investigated two types of energy–restricted diets (55% of energy from carbohydrates, 10% from MUFAs and 40% of energy from carbohydrates, 25% from MUFAs) in longitudinal clinical intervention study among 32 overweight subjects with BMI of 26-45. The MUFA-rich diet showed better effects on serum triglycerides than the carbohydrates-rich diet, even with energy restriction and weight loss [74]. Moreover, a diet rich in MUFA may improve glucose tolerance, insulin sensitivity among the patients with Diabetes Mellitus and is resistant to oxidative modification. The effects of MUFA was also studied in the DELTA study which replaced saturated fats with MUFA and was reported to raise HDL without affecting LDL with decrease in total cholesterol and triglycerides [75].

The major non-animal sources of monounsaturated fat include olive oil, canola oil, nuts and avocados whereas beef is a major animal source for MUFA in western diet. Monounsaturated fats are also typically high in vitamin E which is usually low in western diets. Oleic acid, a cis –isomer, is the major type of monounsaturated fatty acid present in olive oil, nuts and avocados. All studies have shown that oleic acid lowers the plasma cholesterol when substituted for palmitic acid.

Trans Fatty Acid and CHD Risk

Trans fatty acids are the trans isomer of monounsaturated fatty acid that have shown to adversely affect the plasma lipids and lipoproteins. In numerous metabolic controlled metabolic studies, trans fatty acids have shown to raise LDL-C levels and lower HDL-C relative to cis-unsaturated fatty acids, and the increase in the ratio of total to HDL cholesterol for trans –fat to approximately twice that for saturated fat [76-84]. Trans-fat increases plasma levels of lipoprotein a, and triglycerides and may reduce endothelial functions by impairing the flow-mediated dilation [85-86]. In addition, trans-fatty acids adversely affect essential fatty acid metabolism and prostaglandin balance by inhibiting the enzyme delta-6 desaturase [87]. Finally, a high intake of trans-fat may promote insulin resistance and increase risk of type 2 diabetes [88]. The major sources of trans fatty acids in the U.S are stick margarine, commercially baked potatoes, deep fried fast foods and vegetable shortenings.

The positive relationship between trans fatty acid and the risk for CHD have been established in numerous studies. In the Nurses Health Study, the replacement of two percent of energy from trans fat with monounsaturated fats would reduce CHD risk by 53 percent ($p<0.001$). These finding are consistent with results of three large prospective studies which have found elevated risk of CHD to be associated with higher trans fatty acid intake. Oomen et al combined the results from these four prospective studies and reported that pooled relative risk of CHD associated with a difference of 2% energy in trans-fatty acid intake assessed at a baseline of 1.25 (95% CI 1.11-1.40) [89]. Similarly, in a population-based case control study of 179 cardiac arrest patients and 285 community controls, higher red –cell membrane levels of trans-fatty acids, especially trans isomer from partially hydrogenated vegetable oils, were associated with significantly increased risk of primary cardiac arrest. Conversely, results from a small UK study on sudden death, and case control studies using biochemical markers of trans fat intake, have shown no association or have been less consistent respectively [90].

Polyunsaturated Fats and CHD Risk

Two types of polyunsaturated fatty acids (PUFA) occur in diet, n-6 and n-3 polyunsaturates. Linoleic acid (18:2) is the predominant n-6 type of polyunsaturated fatty acid that comes mainly from plant oils. Linolenic acid (18:3) is a parent n-3 polyunsatuarted that also occurs in certain vegetable oils. Soybean oil, rapeseed oil and linseed oil are particularly rich in linolenic acid. In addition to the vegetable source, fish oil also contains large amount of a very long chain of linolenic acid of plant sources. A major type of n-3 fatty acid in fish oil are eicosapentaenoic acid (EPA) 20:5, n-3 and decosahexanoic acid (DHA) 22:6,n-3. Together these constitute for 26% of fatty acid in fish oil.

Linoleic Acid- N-6 PUFA

Existing evidence from many years suggests that linoleic acid has cholesterol lowering effects. Earlier work of Kinsell et al and Ahrens et al showed that vegetable oils rich in linoleic acid lower the serum cholesterol when replaced for dietary saturated fatty acid [91-94]. These findings were confirmed in subsequent studies done by Keys and Hegsted [3-4] who reported that linoleic acid lowers the serum cholesterol more than oleic acid or carbohydrates. In addition to these favorable plasma lipid changes by linoleic acid, many prospective cohort studies illustrated a strong to borderline inverse association between CHD and n-6 PUFA intake. Among these were Western electrical study, Ireland Boston Heart Study [95] and the control group of the multiple risk factor intervention trial. Similarly, in the Nurse Health Study there was an inverse association between PUFA and CHD observed, which was even stronger than previously reported. In animal and metabolic studies, an increased consumption of n-6 PUFA was associated with improved insulin sensitivity and suggested to have anti-arrhythmic effects.

In the Nurses Health Study, the ratio of polyunsaturated fat to long chain saturated fatty acids (P: S ratio) was strongly associated with a lower risk of CHD. After the adjustment of other confounding factors, the RR of CHD was 0.79 (95% CI 0.70-0.89) for each 0.2 unit increment in the ratio. The data suggests that replacing long chain saturated fats with PUFA substantially decreases the risk of CHD. In addition, when intakes of PUFA and trans fat were considered together, the observed risk for CHD was low in the lowest quintile of trans fat and highest quintile of PUFA.

Alpha Linolenic Acid-N-3 PUFA

Alpha Linolenic acid (ALA), n-3 PUFA, found in high concentration in flaxseed, canola and soyabean oils can be metabolized to long chains of EPA and DHA in humans through desaturation – chain elongation pathway. ALA is an essential N-3 PUFA for humans needed in adequate amounts. It is especially important for infants, young children and patients requiring parenteral and enteral nutrition. ALA may have antiarrhythmic effects, lowers serum triglycerides levels, decreases thrombotic tendency, and improves endothelial function.

Antiarrhythmic effects of ALA have been established in many experimental studies. In animal experiments in which arrhythmias were induced by coronary occlusion and reperfusion, significant reductions in the incidence of ventricular fibrillation and cardiac mortality were observed in rats fed on ALA-rich diet. The antiarrythmic effects of ALA may be attributed to an increased electrical threshold for the induction of ventricular fibrillation.

An inverse association between the intake of linolenic acid and risk of fatal CHD was observed in most prospective studies, but not in one smaller study. In the usual care group of multiple risk factor intervention trial, men in the highest quintile of ALA intake had 40% lower CHD mortality compared to men in lowest quintile [96]. Likewise, in the Health Professional Follow-up Study, an increase of 1% in ALA intake was associated with 40% lower risk of fatal CHD. Most recent evidence of the inverse relationship of ALA with CHD comes from the Nurses' Health Study during the 10 years follow-up. After adjusting for the coronary risk factors, RRs of fatal CHD from lowest to highest quintiles of ALA intake were 1.0, 0.89, 0.90, 0.66 and 0.52 (95% CI 0.30-0.90). The major contributors of ALA in the Nurses Health study were oil and vinegar salad dressing. In fact, in a cohort of women, frequent consumption of oil and vinegar salad dressing (source of linolenic acid in US diets) was associated with significantly 50% lower risk of fatal CHD. The effect of ALA supplementation in CHD prevention has also been found in intervention studies. The Indian Experiment of Infarct Survival[97] has reported a 19% reduction in a group assigned to mustard-seed oil (20g/d mustard-seed oil provide 2.9g of ALA). In the Lyon diet heart study ALA-rich Mediterranean diet [98-99], the subjects in the intervention group had a 50-70% reduction in the cardiac end points (cardiac deaths, non fatal MI, unstable angina, heart failure, pulmonary or peripheral embolism, hospital admissions). Similarly, the Indo-Mediterranean Diet Heart Study [100] randomized 1000 subjects with angina pectoris, MI or CAD risk factors in a single blinded trial. The intervention group, who consumed a diet rich in ALA, had fewer total cardiac end points than the control group after two years. In contrast to this, there are two negative studies which reported no significant difference between the intervention and control group. These conflicting results from the studies indicate that further, well-designed trials are needed to determine the exact role of ALA in altering CHD risk. Nevertheless, on the basis of current evidence that support favorable effects of ALA on cardiovascular disease, the inclusion of ALA-rich food along with a low saturated fat, low carbohydrate diet may be advantageous for the prevention of heart disease.

Fish Oil – N-3 PUFA

Fish oil, which is also rich in N-3 PUFA (EPA +DHA), has been shown to lower the risk for CHD in numerous prospective and controlled clinical trials. Kromhout et al first reported that the intake of fish oil has an inverse relationship with coronary mortality in the Dutch population of the Seven Countries study [101]. A 50% lower CHD mortality among men who consume 30g of fish per day was reported than men who rarely ate fish. Similar association was reported in more than 15 following prospective studies [102]. Additional evidence of the protective effects of fish oil against CHD comes from another 4 recent prospective cohort studies; and one case control study done in a diverse population [103]. In fact, in two of the recent studies, it was shown that consuming 2 or more servings of fish per day resulted in a 30% lower risk of CHD in women and that blood levels of EPA and DHA were strongly associated with a decreased risk of sudden death in men [104-106]. Three clinical trials have examined the effects of N-3 PUFA in the secondary prevention of CHD. Burr et al examined the effects of an increased fatty fish intake on the secondary prevention of CHD in diet and reinfarction study (DART). A 29 % reduction in all-cause mortality was found in male MI survivors who were advised to increase fish intake over a two year period of time. In the Indian Experiment of Infarct Survival, as discussed earlier, the effects of fish oil capsules (1.8 g EPA+DHA/d), 20g/d mustard-seed oil providing 2.9 g of linolenic acid/d or placebo were determined. At a follow-up of 1 year, it was shown that total cardiac events were reduced by 30% in the fish oil group with a 19% reduction in the mustard oil group. Similarly, in the GISSI Prevention Trial, 11,324 patients with pre-existing CHD were randomized to 300 mg Vitamin E, fish oil (850mg EPA+DHA) , both vitamin E and fish oil, or neither. At the 3 year follow-up, a 15 % reduction in primary end point of death, non-fatal MI and non fatal stroke($p<0.02$) was noted in fish oil cohort [107-108]. Most of the reduction was attributable to the decrease in cardiovascular disease, especially sudden cardiac death. These clinical trials provide support for the possible role of fish oil in secondary prevention. However, the evidence for using fish oil for primary prevention is still not sufficient and continues to be a topic of discussion.

Carbohydrates, Plasma Lipids and Lipoproteins

Beginning in the early 1970's, carbohydrates were recommended as the preferred substitute for the fat in diet by the American Heart Association and others to achieve successive reduction in total fat and low density lipoprotein cholesterol. Since then, the consumption of carbohydrates in the United States has been increased from 42.4 % kcal to 49.0% for men and from 44.8% kcal to 49.7% kcal for women [109]. This increase has been also accompanied by the increase in energy intake from 2450 kcal to 2618 kcal for men and 1542 kcal to 1877 kcal for women ($p<0.01$), and a parallel decrease in the percentage of calories from total fat (36.9% to 32.8% for men and 36.8% for women ($p<0.01$) [109]. This amount of carbohydrates intake is well within the range of recommended current dietary guidelines for carbohydrate intake. The institute of Medicine of the National Academics recommends a diet that provides 45% to 65% kcal for carbohydrates and 20% to 35% kcal from total fat, while kcal from saturated fats, trans fats and dietary cholesterol are kept low [110]. The American Heart Association and National Cholesterol Education Program also recommend a similar diet that provides 50-60% kcal from carbohydrates. These guidelines are consistent with the dietary guidelines for Americans 2005 [111]. Following these guidelines, the consumption of carbohydrates is on the rise and more Americans consume their daily calories from carbohydrates than ever before. However, much less attention has been given to the expected degree of benefit in terms of carbohydrates effects on plasma lipids and lipoproteins, and associated CHD risk. It has been universally documented that substituting saturated fats with carbohydrates reduces total and LDL cholesterol [112-115]. However, overwhelming available data suggests that the lowering of total cholesterol is partly secondary to the fall in HDL-C. That a highcarbohydrates diet lowers HDL-C was first reported in 1966 by Levy et al and has been a consistent finding in all the subsequent trials [116]. Combined data from 27 well-controlled trials which studies 682 patients in total found that HDL cholesterol decreases by 0.5mg/dl for every 1% energy replaced from saturated fatty acids by carbohydrates [117]. Similar findings were reported in the Meta analyses by Hegsted et al and by Yu et al [118-119]. Thus, indicating that although the inverse relationship between the dietary carbohydrates and total cholesterol and LDL-C may appear beneficial, the overall impact on plasma lipids and lipoproteins is still unfavorable because of a proportionally greater decrease in HDL-C, resulting in a relative increase in the total cholesterol: HDL-C ratio.

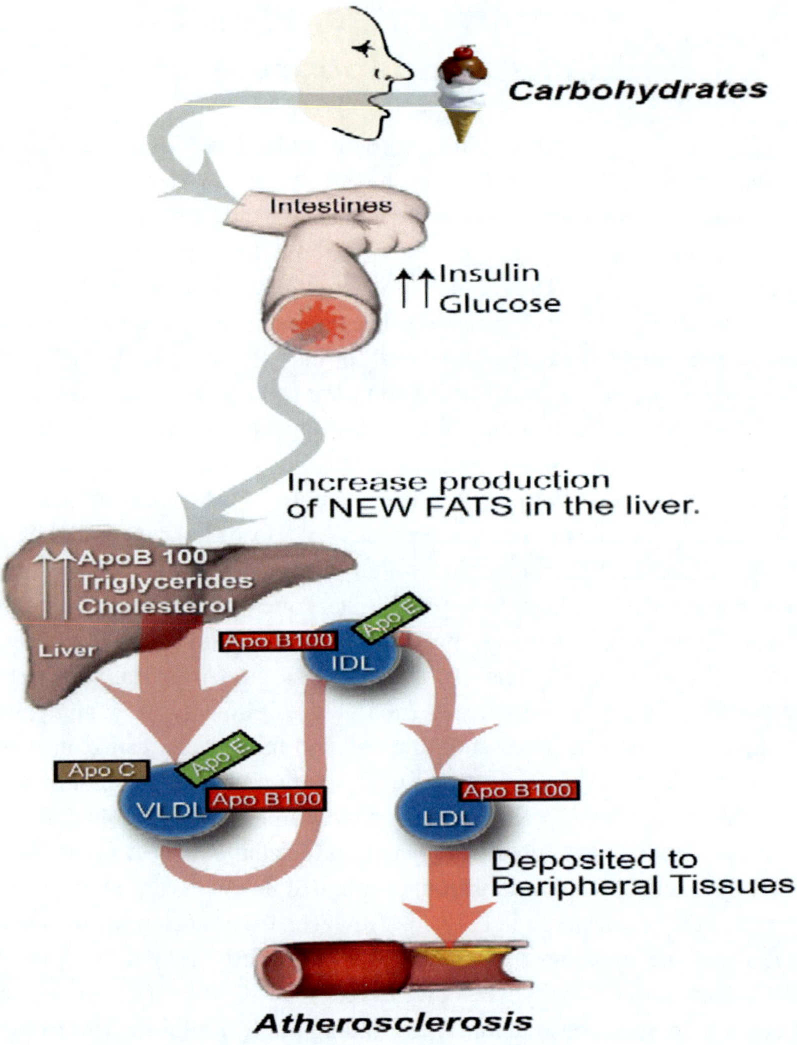

Figure 5. Carbohydrates & De Novo Lipogenesis

In addition to the adverse effect of a fat restricted/ high carbohydrate diet on HDL-C, these diets have the propensity to increase fasting triglyceride levels. Four of five epidemiological cross-sectional analyses including the Nurses Health Study and examination of NHANESIII data demonstrated lower HDL-C and higher triglycerides levels among the individual with the highest intake of carbohydrates, even after the adjustment of confounding factors [120-122]. In a study that looked at the acute effects of carbohydrate feeding on post-prandial triglyceride metabolism reported that the addition of sucrose to a fat rich meal increases the post-prandial triglycerides compared with a test meal containing fats only [123]. With the same line of thinking, Chen et al showed that a very low fat high carbohydrate diet (25% and 60% dietary energy from fat and carbohydrates respectively) for 2 weeks significantly increased post-prandial TG levels [124] Even at a more realistic dietary fat level (30% and 55% from fats and carbohydrates respectively), both fasting and post-prandial TG concentrations were increased as compared to those in high fat diet (45% energy from diet). On the other hand, recently, in a Meta analysis of 8 447 subjects from six randomized controlled trials reported negligible effects of low–carbohydrates on weight loss and other cardiovascular risk factors. Low carbohydrates diets were associated with the favorable changes in HDL-C (4.6mg/dl, 95% CI, 1.5-8.1 mg/dl; p=0.01) and triglycerides (-22.1 mg/dl; 95% CI, -38.1 to -5.3 mg/dl; p=0.13). However, the associated changes in total cholesterol and LDL-C were unfavorably linked to low carbohydrate diets [125].

Furthermore, there is compelling evidence that carbohydrate-rich diets have shown to induce atherogenic triglyceride rich lipoproteins (TRL) and small dense LDL particles which have been implicated as independent risk factors for CHD. TRL is a heterogeneous group of chylomicrons (Apo B 48), large VLDL particles (ApoB100), and chylomicrons remnants. This induction of TRL by a high carbohydrate diet is a result of de novo lipogenesis secondary to increase in the flux of either glucose or fructose along the glycolytic pathway, thereby increasing the availability of acetyl CoA and stimulating fatty acid synthesis [126-128](Figure5). Also, increased carbohydrate intake up regulates the activity of enzymes needed for fatty acid synthesis leading to an increase of TRL secretion. An analysis done by Krauss on several short-term feeding studies indicate that across a wide range of dietary carbohydrates (40%-80%kcal) and fat (5%-45% kcal) intake, there is a strong linear relationship between the prevalence of small dense LDL particles (LDL phenotype B) and the percentage of calories from dietary carbohydrates (r=0.93; p <0.0001) [129]. Moreover, these atherogenic changes appear to be more pronounced in sedentary, overweight or obese populations and the extent of these changes is dependent on the magnitude of the shift from fat to carbohydrates. Similarly, in another study by Krauss et al showed that while low-fat high carbohydrate diet significantly reduced LDL cholesterol concentration, this dietary change also caused 44% of LDL pattern A (large LDL particles) subjects to convert to LDL pattern B (small dense LDL particles) [129]. In another 12 week weight loss intervention study, carbohydrates restriction favorably alters plasma lipoprotein by reducing plasma LDL-C, TG and increasing LDL particle size [130]. All of this evidence suggests that there is substantial evidence that high carbohydrates diets adversely affect the plasma lipids and lipoproteins.

Carbohydrates and Metabolic Syndrome

In the United States, approximately 47 million people have metabolic syndrome (MetS). It is estimated that soon, MetS will be the number one risk factor for cardiovascular disease risk among US populations. Metabolic syndrome represents a constellation of markers that indicates a predisposition to diabetes, cardiovascular disease and other pathologic states. Exact definition and treatment for MetS are a matter of current debate and there is not general consensus available. The National Cholesterol Education Adult Treatment Panel III [131] defined MetS as the presence of three or more of the following 5 factors: (1) central obesity (waist circumference >102 cm for males and >88 cm for females); (2) elevated serum triglycerides (≥ 150mg/dl); (3) low HDL cholesterol (<40mg/dl in men, < 50mg/dl in women); (4) hypertension (≥130/85 mm Hg) and elevated fasting blood glucose(> 110mg/dl). In addition to these lipid abnormalities, small dense LDL particles are also present in the metabolic syndrome (Tab.1). As evident from the above mentioned literature, low carbohydrate diets have a propensity to ameliorate the high plasma TG, low HDL, high blood pressure, high fasting glucose, insulin resistance and small dense LDL particles. The numerous single arm studies analyzed the effects of carbohydrate-restricted diets (ranging from 6-50g/day) of different durations (2 weeks –8 weeks) in subjects that were overweight, presented with symptoms of MetS or were diabetic[132-134]. The results clearly demonstrated that a carbohydrate restricted diet was effective in improving all five parameters of MetS. Similarly, a study done by Garg et al at a four center, randomized study of patients with type II diabetes receiving the glipizide treatment compared diets where MUFA was substituted for carbohydrates or vice a versa [135]. The study confirms that high carbohydrate diets increase plasma TG levels and VLDL-C concentrations (MetS criteria) in NIDDM patients by 24% and 23% respectively compared with monounsaturated fat diets. Consistent with other studies results, this study also reported that high carbohydrate diets offer no advantage in lowering LDL cholesterol levels compared with high fat diets that are low in saturated fats.

Table 1. AHA Criteria for Metabolic syndrome

Criteria for Metabolic Syndrome *American Heart Association and the National Heart, Lung, and Blood Institute*
Elevated waist circumference: Men — Equal to or greater than 40 inches (102 cm) Women — Equal to or greater than 35 inches (88 cm)
Elevated triglycerides: Equal to or greater than 150 mg/dL
Reduced HDL ("good") cholesterol: Men — Less than 40 mg/dL Women — Less than 50 mg/dL
Elevated blood pressure: Equal to or greater than 130/85 mm Hg
Elevated fasting glucose: Equal to or greater than 100 mg/dL

The importance of a LDL subpopulation has been recently increasingly appreciated in respect to its potential atherogenic role and, unlike total LDL, changes in specific LDL particles show a consistent pattern with dietary changes. As discussed earlier, carbohydrates are the major dietary factor of the expression of type B LDL pattern. Published data suggests that type B phenotype rarely occurs in isolation and is highly linked and co-expressed with other features of MetS (especially elevated TG and low HDL). Krauss and colleagues reported that switching from a low carbohydrate/high fat diet (46%) to high carbohydrate/low fat diet (26%) resulted in the worsening of TG and HDL and simultaneous conversion of pattern A to pattern B. Furthermore, carbohydrate restriction effectively and reliably reduces the TG/HDL ratio (two important components of MetS), which is considered a good marker for insulin resistance and risk of cardiovascular disease suggested by a number of recent studies. McLaughlin et al who examined the correlation between insulin resistance, insulin and octreotide. They concluded that TG and HDL were independently related to the insulin resistance and TG/HDL ratio was the best predictor of insulin resistance [136]. In fact, this study was the only study among various other recently reported studies that have shown correspondence between TG/HDL and insulin resistance and CVD risk as measured by LDL particle size. Conversely, other authors have argued that the beneficial effects of low carbohydrate diets are ameliorated by concomitant weight loss achieved on low carbohydrate diets. Volek et al [137] specifically looked at this matter in normal weight men and women who maintained their weight during a six week study comparing low carbohydrate diet (46g/day) with low fat high carbohydrate diet (271g/day). This study reported an improvement in HDL and TG by 11.5% and -40% respectively. Additionally, there was a 40% improvement in TG/HDL; and 35 % improvement in insulin levels. Allick et al[138] reported similar findings in a 2 week study among type II diabetics patient with no weight loss (TG reduced by 44%, TG/HDL by 55% , insulin by 17% where as HDL increased 23.5%). Thus, these findings indicate that overall, the low carbohydrates regimen clearly improves MetS, insulin resistance and LDL particle size.

Carbohydrates and Cardiovascular Disease Risk

Carbohydrate-restricted diet regimens as discussed above are shown to have a significant positive effect in body weight reduction, fasting and post-prandial triglycerides, HDL cholesterol, LDL particle size, Apo B response, and in overall improvement of MetS. However, itis important to know whether such decreases in the different markers on carbohydrate restricted diet correspond to a concomitant decrease in the risk of coronary heart disease in the published data. The study done by Hu et al estimated the risk of CHD associated with isocaloric substitutions of different macronutrients. They reported that an isocaloric substitution of five percent energy from MUFA and PUFA with carbohydrates results in an increase in estimated risk of CHD by 25 % and 60% respectively. Similarly, according to the Framingham study, for each 1 mg/dl decrease of HDL-C, CHD risks increases by 3% in women and 2% in men. Based on the results from this study, 1.6mg/dl decrease in HDL –C was in turn related to a 5.1 % increase in CHD risk [139]. The Quebec cardiovascular heart study reported a twenty fold increase in cardiovascular disease risk with an increase in fasting insulin, Apo B and small dense LDL particles [140]. In the same way, recent epidemiological, prospective, and clinical information suggests the relative importance of TG as a CHD risk factor. A Meta analysis of seventeen population-based prospective studies showed TG as an independent risk factor for CHD [141]. This analysis showed that an 88.5mg/dl (1mmol/l) increase in TG concentrations increased the cardiovascular risk by 32% in men and 76% in women. The prospective physician's heart health study demonstrated that non-fasting serum TG was a significant predictor of future myocardial infarction [142].

Aside from lipid abnormalities, hyperglycemia and hyperinsulinemia, high carbohydrate diets may also lead to hypertension, impaired fibrinolysis, as well as endothelial and inflammatory responses are predictive of increased CHD risk. Few prospective studies have directly related carbohydrate intake and incidence of CHD. In the Honululu Heart study, McGee and colleagues documented 456 cases of CHD among 8,000 men of Japanese ancestry who were followed for 10 years. Measuring the diets at the baseline with a single 24 hour recall, the researchers reported that the intake of calories and carbohydrates was lower among men who later developed CHD than was the rest of cohort participants. However, no statistically significant associations were found in multivariate analysis when both calories and carbohydrates were included in the same model. In the Puerto Rico Heart Health program, researchers established an inverse relationship between carbohydrates intake and CHD. However, the study was not adjusted for total energy intake and increased physical activity (a known protective factor for CHD). Although there are only few prospective studies which investigated the direct relationship of carbohydrates and CHD risk, the association of high-carbohydrates diets with a heightened degree of dyslipidemia, insulin resistance, obesity and metabolic syndrome raises doubts on the recommended low fat/ high carbohydrate diets for the prevention of coronary heart disease.

Areas of Uncertainty

From available literature it appears that the low carbohydrate diets tend to reduce triglycerides, raise HDL-C, increase LDL particle size, improves glycemic control and insulin resistance. However, there are numerous gray areas that are still needed to be explored. First, how low is to be considered a low carbohydrate diet? Published data suggests that some degree of carbohydrate restriction possibly would provide a first line of attack against atherogenic dyslipidemia, symptoms of MetS and insulin resistance. This notion is specifically based on the fact that carbohydrate restriction will inhibit the insulin trigger, keep insulin level low, and as a result, the body's metabolism will be geared towards the lipids oxidation rather than storage. Thus, it implies that probably lower is better in terms of carbohydrate consumption and indicates, for example, that a 40% carbohydrate diet would be better than a 55% carbohydrate diet for overall improvement of dyslipidemia and MetS. This idea of "the lower the better" in terms of carbohydrate is at least supported by the work of Krauss who demonstrated that LDL pattern type B associated with MetS and insulin, showed an inverse correspondence with carbohydrate intakes. Secondly, the issue of carbohydrate related factors (glycemic index or glycemic load) in terms of keeping insulin triggers at a minimum needs to be investigated further. Glycemic index is a ranking of foods containing carbohydrates according to their glycemic effects compared to either white bread or glucose. No doubt that higher glycemic food will trigger insulin excessively, but does the low glycemic index (low sugar, complex carbohydrates) really inhibit insulin triggers, keeps insulin level under control and prevent accumulation of atherogenic dyslipidemic LDL particles and markers of MetS? Moreover, studies demonstrated that the adverse effect of carbohydrate was not dependent on whether that carbohydrate was rich in dietary fiber or low in glycemic index [143-145]. The fact that a low-glycemic index or high fiber diet can exert beneficial effects on blood glucose; data from their long term effect on insulin sensitivity are unconvincing [146-152]. Thus, indicating that segregating carbohydrates on the basis of glycemic load or index needs more evidence. Finally, the effect of low carbohydrate intake on long-term impact of carbohydrate intake on CHD risk is still controversial and less clear. Meta analysis of numerous randomized control trials for low carbohydrate diets indicated that no study reported an effect of a low carbohydrate diet on CHD mortality or morbidity irrespective of marked improvement in different biomarkers associated with CHD. The probable explanation is the presence of other confounding factors present in these trials. Thus, clearly pointing out that long term intervention studies in diverse population are needed to provide more statistical power to the relationship of carbohydrates with CHD morbidity and mortality.

Conclusion

Taken together, compelling evidence indicates the greater importance of different types of macronutrients with respect to the risk of coronary heart disease. The evidence that dietary cholesterol is associated with CHD risk is not conclusive. The strong inter-relation between dietary and serum cholesterol found in many species has not been replicated among humans, and is rather attributed to certain eating patterns high in saturated fats and trans fats. Numerous epidemiological, clinical trials, and meta analysis in the past several decades indicate that substituting saturated fats and trans fatty acid with poly and mono unsaturated fats is more effective in positively altering the plasma lipids and lipoproteins, as well as reducing the risk of coronary heart disease. Saturated fats also have deleterious effects on endothelial cell lining, promote vascular inflammation and decrease insulin sensitivity. It is apparent from available scientific data that widely promoted low fat regimen is based on little scientific and dietary total fats are not related to increased cardiovascular risk. Conversely, the optimal amount and mixture of mono and poly unsaturated fats that should be used in the diet still remain unsettled. The scientific data suggests beneficial effects of both N-6 PUFA and N-3 PUFA (plant and marine derived) on reducing CHD risk. There are some concerns about the optimal balance between N-3 PUFA and N-6 PUFA daily consumptions. Some researchers have proposed that reducing the N-6 PUFA to achieve greater ratio of N-3PUFA to N-6 PUFA. However, there is little evidence that higher ratio predicts additional benefits that lower CHD risk [153].

Additionally, the concept of low carbohydrate diets continues to emerge, evidence is mounting to support low carbohydrate diets as an effective means for the improvement of cardiovascular risk factor and overall CHD risk. Published data suggests that low carbohydrate diets have out-performed recommended low fat diets in improving blood glucose and triglycerides level as well as raising HDL cholesterol. Moreover, low carbohydrate diets are possibly the most important strategy that positively modifies all parameters of metabolic syndrome and insulin resistance. The basis of this idea is that a low carbohydrate diet results in the disruption of insulin release and metabolism. However, one should also remember that it is difficult to isolate the independent effects of individual macronutrients on plasma lipids. The available data is confounded by the changes in body weight and presence of two or more macronutrients in a single type of food. Also, there is high degree of variability in response among individuals to different macronutrients. Nonetheless, scientific data from numerous studies converges to indicate that at least the dietary strategy of substituting saturated fats and trans fats with more unsaturated fats; increasing the consumption of more N-3 PUFA from both fish and plant source along with lowering carbohydrates would be effective in favorably altering cardiovascular disease risk.

References

[1] Quintao E, Grundy SM, Ahrens EH Jr: Effects of dietary cholesterol on the regulation of total body cholesterol in man. *J. Lipid Res*. 1971.12: 233-247.

[2] Grundy SM: Absorption and metabolism of dietary cholesterol. *Annu. Rev. Nutz* 1983 3: 71-96.

[3] Keys A., Anderson JT, Grande F: Serum cholesterol response to changes in the diet. IV. Particular saturated fatty acids in the diet. *Metabolism.* 1965;14: 776-787.

[4] Hegsted DM, McGandy RB, Myers ML, Stare EJ: Quantitative effects of dietary fat on serum cholesterol in man. *Am. J. Clin. Nutr.* 1965. 17: 281-295.

[5] Strong JP, McGill H. C: Diet and experimental atherosclerosis in baboons. *Am. J. Pathol.* 1967. 50:669-690

[6] Armstrong ML: Regression of atherosclerosis. *Atheroscler Rev.* 1975. 1: 137-188.

[7] Strong JP: Atherosclerosis in primates. Introduction and overview. *Primates Med.* 1976. *9:*1-15.

[8] Ginsberg H, Le NA., Mays C, Gibson J, Brown WV: Lipoprotein metabolism in non-responders to increased dietary cholesterol. *Arteriosclerosis*.1981;1: 463-470.

[9] Packard CJ, McKinney L, Carr K, Shepherd C: Cholesterol feeding increases low density lipoprotein synthesis. *J. Clin. Invest.* 1983. 72: 45-51.

[10] Slater G, Mead J, Dhopeshwarkar G, Robinson S, Alfin-Slater RB: Plasma cholesterol and triglycerides in men with added eggs in the diet. *Nutr. Rep. Znt*. 1976;14:249-260

[11] Porter MW, Yamanaka W, Carlson SD, Flynn MA :Effects of dietary egg on serum cholesterol and triglycerides in human males. *Am. J Clin. Nut* 1977; 30*:* 490-495.

[12] Flynn MA., Nolph GB, Flynn TC, Kahns R, Krause G: Effect of dietary egg on human serum cholesterol and triglyceride. *Am. J Clin. Nut.* 1979. 32:1050-1057

[13] Bassett DR, Abel M, Moellering RC Jr, Rosenblatt G, Stokes J III: Coronary heart disease in Hawaii: dietary intake, depot fat, stress, smoking, and energy balance in Hawaiian and Japanese men. *Am J Clin Nutr*.1969;22:1483–1503

[14] Esrey KL, Joseph L, Grover SA: Relationship between dietary intake and coronary heart disease mortality: lipid research clinics prevalence follow-up study. *J Clin Epidemiol* 1996; 49:211–216

[15] Finegan A, Hickey N, Maurer B, Mulcahy R: Diet and coronary heart disease: dietary analysis on 100 male patients. *Am J Clin Nutr* 1968;21:143–148

[16] Garcia-Palmieri MR, Sorlie P, Tillotson J, Costas R Jr, Cordero E, Rodriguez M. Relationship of dietary intake to subsequent coronary heart

disease incidence: The Puerto Rico Heart Health Program. *Am J Clin Nutr.*1980;33:1818–1827

[17] Gordon T, Kagan A, Garcia-Palmieri M ,Kannel WB, Zukel WJ, Tillotson J, Sorlie P, Hjortland M: Diet and its relation to coronary heart disease and death in three populations. *Circulation* 1981; 63:500–515

[18] Hu FB, Stampfer MJ, Manson JE, Rimm E, Colditz GA, Rosner BA, Hennekens CH, Willett WC :Dietary fat intake and the risk of coronary heart disease in women. *N Engl J Med.*1997;337:1491–1499

[19] Khaw KT, Barrett-Connor E: Dietary fiber and reduced ischemic heart disease mortality rates in men and women: a 12-year prospective study.*AmJ Epidemiol.*1987; 126:1093–1102

[20] Morris JN, Marr JW, Clayton DG: Diet and heart: a postscript. *BMJ.*1977;2:1307–1314

[21] Pietinen P, Ascherio A, Korhonen P, Hartman AM, Willett WC, Albanes D, Virtamo J :Intake of fatty acids and risk of coronary heart disease in a cohort of Finnish men. The Alpha-Tocopherol,Beta-Carotene Cancer PreventionStudy.*AmJEpidemiol.* 1997; 145:876–887

[22] Posner BM, Cobb JL, Belanger AJ, Cupples LA, d'Agostino RB, Stokes J III :Dietary lipid predictors of coronary heart disease in men. The Framingham Study.*Arch Intern Med* 1991;151:1181–1187

[23] Yano K, Rhoads GG, Kagan A, Tillotson J: Dietary intake and the risk of coronary heart disease in Japanese men living in Hawaii. *Am J Clin Nutr* 1978;31:1270–1279

[24] Katan MB, Beynen AC, de Vries JH, Nobels: A Existence of consistent hypo- and hyper-responders to dietary cholesterol in man. *Am J Epidemiol.*1986; 123:221–234

[25] Katan MB, Berns MA, Glatz JF, Knuiman JT, Nobels A, de Vries JH: Congruence of individual responsiveness to dietary cholesterol and to saturated fat in humans. *J Lipid Res* 1988;29:883–892

[26] Bosner MS, Lange LG, Stenson WF, Ostlund RE, Jr: Percent of cholesterol absorption in normal women and men quantified with dual stable isotopic tracers and negative ion mass spectrometry. *J Lipid Res.* 1999;40:302–308

[27] Ostlund RE Jr, Bosner MS, Stenson WF: Cholesterol absorption efficiency declines at moderate dietary doses in normal human subjects. *J Lipid Res.*1999;40:1453–1458

[28] Shekelle RB, Shryock AM, Paul O, Lepper M, Stamler J, Liu S, Raynor WJ Jr : Diet, serum cholesterol, and death from coronary heart disease. The Western Electric study. *N Engl J Med.*1981; 304:65–70

[29] Shekelle RB, Stamler J: Dietary cholesterol and ischemic heart disease. *Lancet*.1989;1:1177– 1179

[30] Kromhout D, de Lezene Coulander C: Diet, prevalence and 10-year mortality from coronary heart disease in 871 middle-aged men. The Zutphen Study. *Am J Epidemiol*.1984; 119:733–741

[31] Toeller M, Buyken AE, Heitkamp G, Scherbaum WA, Krans HM, Fuller JH : Associations of fat and cholesterol intake with serum lipid levels and cardiovascular disease: the EURODIAB IDDM *Complications Study. Exp Clin Endocrinol Diabetes*.1999; 107:512–521

[32] Ascherio A, Rimm EB, Giovannucci EL, Spiegelman D, Stampfer M, Willett WC: Dietary fat and risk of coronary heart disease in men: cohort follow-up study in the United States. *BMJ*. 1996; 313:84–90

[33] Keys A*: "Seven Countries: A Multivariate Analysis of Death and Coronary Heart Disease."* Cambridge, MA: Harvard University Press, 1980.

[34] Dawber TR, Nickerson RJ, Brand FN, Pool J: Eggs, serum cholesterol, and coronary heart disease. *Am J Clin Nutr* 1982; 36:617–625

[35] Hu FB, Stampfer MJ, Rimm EB, Manson JE, Ascherio A, Colditz GA, Rosner BA, Spiegelman D, Speizer FE, Sacks FM, Hennekens CH, Willett WC: A prospective study of egg consumption and risk of cardiovascular disease in men and women. *JAMA* (1999) 281:1387– 1394

[36] Matthan NR, Ausman LM, Lichtenstein AH, Jones PJ : Hydrogenated fat consumption affects cholesterol synthesis in moderately hypercholesterolemic women. *J Lipid Res*. 2000 May; 41(5):834-9

[37] Spady DK, Dietschy JM: Dietary saturated triacylglycerols suppress hepatic low density lipoprotein receptors in the hamster. *Pmc. Natl. Acad. Sci. USA 1985*. 82: 4526-4530.

[38] Nicolosi RJ, Stucchi AE, Kowala MC, Hennessy LK, Hegsted DM, Schaefer EJ: Effect of dietary fat saturation and cholesterol on LDL composition and metabolism. *Arteriosclerosis*. 1990. 10: 119-1 2 8.

[39] Sorci-Thomas M, Wilson MD, Johnson EL, Williams DL, Rudel LL: Studies on the expression of genes encoding apolipoproteins B lOO and B48 and the low density lipoprotein receptor in nonhuman primates. *J. Biol. Chem*. 1989. 264: 9039-9045

[40] Spady DK, Dietschy JM: Interactions of aging and dietary fat in the regulation of low density lipoprotein transport in the hamster. *J. Lid Res*. 1989. 3 0 559-569.

[41] Grande F, Anderson JT, Keys A: Comparison of effects of palmitic and stearic acids in the diet on serum cholesterol in man. *American Journal of Clinical Nutrition*.1970;23, 1184-1193

[42] Grundy SM, Denke MA: Dietary influences on serum lipids and lipoproteins. *Journal of Lipid Research* 31, 1990;1149–1172

[43] Watts GF, Jackson P, Burke V, Lewis B : Dietary fatty acids and progression of coronary artery disease in men. *Am J Clin Nutr.* 1996 Aug;64(2):202-9.

[44] 44.Simon JA, Hodgkins ML, Browner WS, Neuhaus JM, Bernert JT Jr, Hulley SB :Serum fatty acids and the risk of coronary heart disease. *Am J Epidemiol.* 1995;Sep 1; 142(5):469-76.

[45] Kabagambe EK, Baylin A, Siles X, Campos H :Individual saturated fatty acids and nonfatal acute myocardial infarction in Costa Rica. *Eur J Clin Nutr.* 2003 Nov; 57(11):1447-57

[46] Ahrens EH, Hirsch J, Insull W, Tsaltas TT, Blomstrand R, Peterson ML :The influence of dietary fats on serum-lipid levels in man. *Lancet.* 1957. 1: 943-953.

[47] Bonanome A, Grundy SM: Effect of dietary stearic acid on plasma cholesterol and lipoprotein levels. *N. Eng J. Med.* 1988;318: 1244-1248.

[48] Aro A, Jauhiainen M, Partanen R, Salminen I, Mutanen M: Stearic acid, trans fatty acids, and dairy fat: effects on serum and lipoprotein lipids, apolipoproteins, lipoprotein(a), and lipid transfer proteins in healthy subjects. *Am J Clin Nutr* 65:1491–1426, 1997.

[49] Frantz ID Jr, Dawson EA, Ashman PL : Test of effect of lipid lowering by diet on cardiovascular risk: the Minnesota Coronary Survey. *Arteriosclerosis.* 1989; 9:129-135

[50] Dayton S, Pearce ML, Hashimoto S, Dixon WJ, Tomiyasu U :A controlled clinical trial of a diet high in unsaturated fat in preventing complications of atherosclerosis. *Circulation.* 1969; 40(suppl 2):1-63.

[51] Turpeinen O, Karvonen MJ, Pekkarinen M, Miettinen M, Elosuo R, Paavilainen E: Dietary prevention of coronary heart disease: the Finnish Mental Hospital Study. *Int J Epidemiol.* 1979; 8:99-118.

[52] Leren P: The Oslo Diet-Heart Study. *Acta Med Scand.* 1966;466(suppl):5-92.

[53] Leren P: The Oslo Diet-Heart Study: eleven-year report. *Circulation.* 1970; 42:935-942.

[54] Morris JN, Ball KP, Antonis A: Controlled trial of soy-bean oil in myocardial infarction. *Lancet.* 1968; 2:693-699

[55] Ball KP, Hanington E, McAllen PM: Low-fat diet in myocardial infarction: a controlled trial. *Lancet.* 1965; 2:501-504.

[56] Burr ML, Fehily AM, Gilbert JF: Effects of changes in fat, fish, and fiber intakes on death and myocardial reinfarction: Diet and Reinfarction Trial (DART). *Lancet.* 1989; 2:757-761.

[57] Nicholls SJ, Lundman P, Harmer JA, Cutri B, Griffiths KA, Rye KA, Barter PJ, Celermajer DS. Consumption of saturated fat impairs the anti-inflammatory properties of high-density lipoproteins and endothelial function. *J Am Coll Cardiol.* 2006 Aug 15; 48(4):715-20. Epub 2006 Jul 24

[58] Raitakari OT, Lai N, Griffiths K, McCredie R, Sullivan D, Celermajer DS: Enhanced peripheral vasodilation in humans after a fatty meal. *J Am Coll Cardiol* 2000;36:417–22.

[59] Plotnick GD, Corretti MC, Vogel RA: Effect of antioxidant vitamins on the transient impairment of endothelium-dependent brachial artery vasoactivity following a single high-fat meal. *JAMA* 1997;278:1682–6.

[60] Vogel RA, Corretti MC, Plotnick GD: Effect of a single high-fat meal on endothelial function in healthy subjects. *Am J Cardiol* 1997;79: 350–4.

[61] Keogh JB, Grieger JA, Noakes M, Clifton PM: Flow-mediated dilatation is impaired by a high-saturated fat diet but not by a high carbohydrate diet. *Arterioscler Thromb Vasc Biol* 2005; 25:1274–9.

[62] Vogel RA, Corretti MC, Plotnick GD: The post-prandial effect of components of the Mediterranean diet on endothelial function. *J Am Coll Cardiol* 2000; 36:1455– 60.

[63] McAuley K, Mann J. Thematic review series: patient-oriented research. Nutritional determinants of insulin resistance *J Lipid Res.* 2006 Aug; 47(8):1668-76. Epub 2006 May 23.

[64] Storlien LH, Higgins JA, Thomas TC, Brown MA, Wang HQ, Huang XF, Else PL. Diet composition and insulin action in animal models. *Br J Nutr.* 2000 Mar; 83 Suppl 1:S85-90.

[65] Clandinin MT, Cheema S, Field CJ, Baracos VE. Dietary lipids *Ann N Y Acad Sci. 1993* Jun 14; 683:151-6. Review

[66] Lönnroth P :Regulation of insulin action at the cellular level._*J Intern Med Suppl.* 1991; 735:23-9.

[67] Manco M, Calvani M, Mingrone G: Effects of dietary fat *Diabetes Obes Metab.* 2004 Nov;6(6):402-13. Review.

[68] Vessby B, Gustafsson IB, Tengblad S, Boberg M, Andersson A: Desaturation and elongation *Ann N Y Acad Sci. 2002* Jun;967:183-95.

[69] 69. Vessby B, Tengblad S, Lithell H: Insulin sensitivity Diabetologia. 1994 Oct; 37(10):1044-50.

[70] Vessby B, Unsitupa M, Hermansen K, Riccardi G, Rivellese AA, Tapsell LC, Nälsén C, Berglund L, Louheranta A, Rasmussen BM, Calvert GD,

Maffetone A, Pedersen E, Gustafsson IB, Storlien LH; KANWU Study. Substituting dietary saturated for monounsaturated fat impairs insulin sensitivity in healthy men and women: The KANWU Study. *Diabetologia.* 2001 Mar; 44(3):312-9

[71] Pérez-Jiménez F, López-Miranda J, Pinillos MD, Gómez P, Paz-Rojas E, Montilla P, Marín C, Velasco MJ, Blanco-Molina A, Jiménez Perepérez JA, Ordovás JM: A Mediterranean and a high-carbohydrate diet improve glucose metabolism in healthy young persons. *Diabetologia.* 2001 Nov;44(11):2038-43.

[72] Paniagua JA, Gallego de la Sacristana A, Romero I, Vidal-Puig A, Latre JM, Sanchez E, Perez-Martinez P, Lopez-Miranda J, Perez-Jimenez F Monounsaturated fat-rich diet prevents central body fat distribution and decreases post-prandial adiponectin expression induced by a carbohydrate-rich diet in insulin-resistant subjects. *Diabetes Care.* 2007 Jul; 30(7):1717-23. Epub 2007 Mar 23.

[73] Colette C, Percheron C, Pares-Herbute N, Michel F, Pham TC, Brillant L, Descomps B, Monnier L : Exchanging carbohydrates for monounsaturated fats in energy-restricted diets: effects on metabolic profile and other cardiovascular risk factors. *Int J Obes Relat Metab Disord.* 2003 Jun; 27(6):648-56

[74] Pietinen P, Ascherio A, Korhonen P, Hartman AM, Willett WC, Albanes D & Virtamo J: Intake of fatty acids and risk of coronary heart disease in a cohort of Finnish men: the Alpha- Tocopherol, Beta-Carotene Cancer Prevention Study. *American Journal of Epidemiology.*1997; 145, 876–887

[75] Ginsberg HN, Kris-Etherton P, Dennis B, Elmer PJ, Ershow A, Lefevre M, Pearson T, Roheim P, Ramakrishnan R, Reed R, Stewart K, Stewart P, Phillips K, Anderson N. Effects of reducing dietary saturated fatty acids on plasma lipids and lipoproteins in healthy subjects: the DELTA Study, protocol 1. *Arterioscler Thromb Vasc Biol.* 1998 Mar; 18(3):441-9

[76] Mensink RP, Katan MB: Effect of dietary trans fatty acids on high-density and low-density lipoprotein cholesterol levels in healthy subjects. *N Engl J Med.* 1990; 323:439-445.

[77] Zock PL, Katan MB : Hydrogenation alternatives: effects of trans fatty acids and stearic acid versus linoleic acid on serum lipids and lipoproteins in humans. *J Lipid Res.* 1992; 33:399-410.

[78] Nestel P, Noakes M, Belling BE : Plasma lipoprotein and Lp[a] changes with substitution of elaidic acid for oleic acid in the diet. *J Lipid Res.* 1992;33:1029-1036.

[79] Judd JT, Clevidence BA, Muesing RA, Wittes J, Sunkin ME, Podczasy JJ: Dietary trans fatty acids: effects of plasma lipids and lipoproteins of healthy men and women. *Am J Clin Nutr*. 1994; 59:861-868

[80] Judd JT, Baer DJ, Clevidence BA, Kris-Etherton P, Muesing RA, Iwane M. Dietary cis? and trans monounsaturated and saturated FA and plasma lipids and lipoproteins in men. *Lipids*. 2002; 37:123-131

[81] Lichtenstein AH, Ausman LM, Jalbert SM, Schaffer EJ: Effects of different forms of dietary hydrogenated fats on serum lipoprotein cholesterol levels. *N Engl J Med*. 1999; 340:1933-1940.

[82] Sundram K, Ismail A, Hays KC, Jeyamalar R, Pathmanathan R : Trans (elaidic) fatty acids adversely affect the lipoprotein profile relative to specific saturated fatty acids in humans. *J Nutr.* 1997;127:514S-520S.

[83] Lichtenstein AH, Ausman LM, Carrasco W, Jenner JL, Ordovas JM, Schaefer EJ: Hydrogenation impairs the hypolipidemic effect of corn oil in humans: hydrogenation, *t*rans fatty acids, and plasma lipids. *Arterioscler Thromb*. 1993; 13:154-161.

[84] Ascherio A, Katan MB, Zock PL, Stampfer MJ, Willett WC: Trans fatty acids and coronary heart disease. *N Engl J Med*. 1999; 340:1994-1998.

[85] de Roos NM, Bots ML, Siebelink E, Schouten E, Katan MB : Flow-mediated vasodilatation is not impaired when HDL-cholesterol is lowered by substituting carbohydrates for monounsaturated fat. *Br J Nutr.* 2001; 86:181-188.

[86] Hill EG, Johnson SB, Lawson LD, Mahfouz MM, Holman RT : Perturbation of the metabolism of essential fatty acids by dietary hydrogenated vegetable oil. *Proc Natl Acad Sci U S A. 1982*; 79:953-957.

[87] Kinsella JE, Bruckner G, Mai J, Shimp J: Metabolism of trans fatty acids with emphasis on the effects of trans, trans-octadecadienoate on lipid composition, essential fatty acid, and prostaglandins: an overview. *Am J Clin Nutr*. 1981; 34:2307-2318

[88] Lovejoy JC. Dietary fatty acids and insulin resistance. *Curr Atheroscler Rep*. 1999; 1:215-220.

[89] Oomen CM, Ocke MC, Feskens EJM, van Erp-Baart M-AJ,Kok FJ & Kromhout D: Association between trans fatty acid intake and 10-year risk of coronary heart disease in the Zutphen Elderly Study: a prospective population-based study. *Lancet*.2001;357, 746–751.

[90] Roberts TL, Wood DA, Riemersma RA, Gallagher PJ, Lampe FC: Trans isomers of oleic and linoleic acids in adipose tissue and sudden cardiac death. *Lancet*. 1995.345:278–282

[91] Kinsell, L. W., J. Partridge, L. Boling, S. Margen, and G. Michaels : Dietary modification of serum cholesterol and phospholipid levels. *J. Clin. Endocrinol*. 1952; 12: 909-913.

[92] Kinsell LW, Michaels GD, Partridge JW, Boling LA, Balch HE, Cochrane GC. Effect upon serum cholesterol and phospholipids of diets containing large amounts of vegetable fat. *J. Clin. Nutr* 1953;1:231-244

[93] Ahrens EH, Hirsch J, Insull W, Tsaltas TT, Blomstrand R, Peterson ML: The influence of dietary fats on serum-lipid levels in man. *Lancet*.1957; 1: 943-953.

[94] Ahrens EH Jr, Blankenhorn DH, Tsaltas TT. Effect on human serum lipids of substituting plant for animal fat in diet. *Proc. Sot. Exp. Biol. Med.* 1954;86:872-878

[95] Kushi LH, Lew RA, Stare FJ, Ellison CR, el Lozy M, Bourke G, Daly L, Graham I, Hickey N, Mulcahy R: Diet and 20-year mortality from coronary heart disease. The Ireland- Boston Diet-Heart Study. *New England Journal of Medicine*.1985; 312, 811–818.

[96] Dolecek TA: Epidemiological evidence of relationships between dietary polyunsaturated fatty acids and mortality in the multiple risk factor intervention trial. *Proceedings of the Society for Experimental Biology and Medicine*.1992; 200, 177–182.

[97] Singh RB, Niaz MA, Sharma JP, Kumar R, Rastogi V, Moshiri M : Randomized, double-blind, placebo controlled trial of fish oil and mustard oil in patients with suspected acute myocardial infarction: the Indian Experiment of Infarct Survival 4. *Cardiovasc Drugs Ther*. 1997; 11:485-491.

[98] de Lorgeril M, Renaud S, Mamelle N: Mediterranean alpha-linolenic acid-rich diet in secondary prevention of coronary heart disease. *Lancet*. 1994; 343:1454-1459.

[99] de Lorgeril M, Salen P, Martin JL, Monjaud I, Delaye J, Mamelle N : Mediterranean diet, traditional risk factors, and the rate of cardiovascular complications after myocardial infarction: final report of the Lyon Diet Heart Study. *Circulation*. 1999;99:779-785.

[100] Singh RB, Dubnov G, Niaz MA, Ghosh S, Singh R, Rastogi SS, Manor O, Pella D, Berry EM: Effect of an Indo-Mediterranean diet on the progression of coronary artery disease in high risk patients (Indo-Mediterranean Diet Heart Study): a randomized single-blind trial. *Lancet*. 2002 Nov 9; 360(9344):1455-61.

[101] Kromhout D, Bosschieter EB & de Lezenne Coulander C: The inverse relation between fish consumption and 20-year mortality from coronary heart disease. New England Journal of Medicine. 1985; 312, 1205–1209.

[102] Oomen CM, Feskens EJ, Rasanen L: Fish consumption and coronary heart disease mortality in Finland, Italy, and the Netherlands.*AmJ Epidemiol.* 2000; 151:999-1006.

[103] Rissanen T, Voutilainen S, Nyyssonen K, Lakka TA, Salonen JT : Fish oil-derived fatty acids, docosahexaenoic acid and docosapentaenoic acid, and the risk of acute coronary events: the Kuopio Ischemic Heart Disease Risk Factor Study. *Circulation*. 2000;102:2677- 2679.

[104] Hu FB, Bronner L, Willett WC: Fish and omega-3 fatty acid and risk of coronary heart disease in women. *JAMA*. 2002; 287:1815-1821.

[105] Albert CM, Campos H, Stampfer MJ, Ridker PM, Manson JE, Willett WC: Blood levels of long chain n-3 fatty acids and the risk of sudden death. *N Engl J Med*. 2002; 346:1113-1118.

[106] Tavani A, Pelucchi C, Negri E, Bertuzzi M, La Vecchia C: n-3 Polyunsaturated fatty acids, fish, and non-fatal acute myocardial infarction. *Circulation*. 2001; 104:2269-2272.

[107] GISSI-Prevenzione Investigators. Dietary supplementation with n-3 polyunsaturated fatty acids and vitamin E after myocardial infarction: results from the GISSI-Prevenzione trial. *Lancet*. 1999;354:447-455.

[108] Marchioli R, Barzi F, Bomba E: Early protection against sudden death by n-3 polyunsaturated fatty acids after myocardial infarction: timecourse analysis of the results of the Gruppo Italiano per lo Studio della Sopravvivenza nell'Infarto Miocardico (GISSI)-Prevenzione. *Circulation*. 2002;105: 1897-1903

[109] Centers for Disease Control. Trends in Intake of Energy and Macronutrients— United States, 1971–2000 *MMWR*. February 6, 2004.

[110] National Academy of Sciences and the Institutes of Medicine., *Dietary Reference Intakes: energy, carbohydrate, fiber, fat, fatty acids, cholesterol, protein, and amino acids*. 2002; Washington, DC

[111] Department of Health and Human Services and the Department of Agriculture. *Dietary Guidelines for Americans*. 2005.

[112] Ginsberg HN, Barr SL, Gilbert A, Karmally W, Deckelbaum R, Kaplan K, Ramakrishnan R, Holleran S, Dell RB: Reduction of plasma cholesterol levels in normal men on an Am Heart Association Step 1 diet or a Step 1 diet with added monounsaturated fat. *N Engl J Med.* 1990;322: 574–579.

[113] Berry EM, Eisenberg S, Friedlander Y, Harats D, Kaufmann NA, Norman Y, Stein Y: Effects of diets rich in monounsaturated fatty acids on plasma

lipoproteins–the Jerusalem Nutrition Study. II Monounsaturated fatty acids vs carbohydrates. *Am J Clin Nutr.* 1992;56:394–403.

[114] Grundy SM, Florentin L, Nix D, Whelan MF: Comparison of monounsaturated fatty acids and carbohydrates for reducing raised levels of plasma cholesterol in man. *Am J Clin Nutr*. 1988;47:965–969.

[115] Jansen S, Lopez-Miranda J, Salas J, Castro P, Paniagua JA, Lopez-Segura F, Ordovas JM, Jimenez-Pereperez JA, Blanco A, Perez-Jimenez F: Plasma lipid response to hypolipidemic diets in young healthy non-obese men varies with body mass index. *J Nutr.* 1998; 128:1144 –1149.

[116] Levy RI, Lees RS, Fredrickson DS: The nature of pre-beta (very low density) lipoproteins. *J Clin Invest* 1966;45:63-77.

[117] Katan MB: High-oil compared with low-fat, high-carbohydrate diets in the prevention of ischemic heart disease *Am J Clin Nutr.* 1997 Oct;66(4 Suppl):974S-979S

[118] Hegsted DM, Ausman LM, Johnson JA, Dallai GE: Dietary fat and serum lipids: an evaluation of the experimental data. *Am J Ciin Nutr* 1993; 57:875-83.

[119] Yu 5, Derr J, Etherton ID, Kris-Etherton PM: Plasma cholesterol predictive equations demonstrate that stearic acid is neutral and monounsaturated fatty acids are hypocholesterolemic. *Am J Clin Nutr.* 1995; 6l: i 129-39.

[120] Buyken AE, Toeller M, Heitkamp G, Karamanos B, Rottiers R, Muggeo M, Fuller JH: Glycemic index in the diet of European outpatients with type 1 diabetes: relations to glycated hemoglobin and serum lipids. *Am J Clin Nutr* 73:574–581, 2001.

[121] Ford ES, Liu S: Glycemic index and serum high-density lipoprotein cholesterol concentration among US adults. *Arch Intern Med* 161:572–576, 2001.

[122] Frost G, Leeds AA, Dore CJ, Madeiros S, Brading S, Dornhorst A: Glycaemic index as a determinant of serum HDL-cholesterol concentration. *Lancet* 353:1045–1048, 1999

[123] Liu S, Manson J, Stampfer M, Holmes M, Hu F, Hankinson S, Willett W: Dietary glycemic load assessed by food-frequency questionnaire in relation to plasma high-density-lipoprotein cholesterol and fasting plasma triacylglycerols in postmenopausal women. *Am J Clin Nutr* 73:560–566, 2001.

[124] Chen YD, Swami S, Skowronski R, Coulston AM, Reaven GM: Effect of variations *J Clin Endocrinol Metab*. 1993 Feb;76(2):347-51

[125] Nordmann AJ, Nordmann A, Briel M, Keller U, Yancy WS Jr, Brehm BJ, Bucher HC: Effects of low-carbohydrate *Arch Intern Med.* 2006 Feb 13;166(3):285-93. Review.

[126] Abbassi F, McLaughlin T, Lamendola C, SunKIm H , Tanaka A, Wang T. High carbohydrate diets, Trigylcerides rich lipoprotein and coronary heart disease. *Am.J cardiol.*

[127] Leary ET, WangT, baker Dj, Cilla DD: Evaluation of immunoseperation method for quantative measurment of remnant like particle-cholesterol in serum and plasma. *Clin Chem* 1998

[128] Jeppesen J, Schaaf P, Jones C, Zhou MY, Chen, YD, Reaven GM: Effects of low-fat, high-carbohydrate diets on risk factors for ischemic heart disease in postmenopausal women *Am. J. Clinical Nutrition*, Apr 1997; 65: 1027 - 1033.

[129] Kruass RM, Dreon DM: Low density lipoprotein subclasses and response to a low –fat diet in healthy men *American Journal of clinical Nutrition* 62, 478S-487S

[130] Miller WM, Nori-Janosz KE, Lillystone M, Yanez J, McCullough PA: Obesity and lipids. *Curr Cardiol Rep*. 2005; 7:465– 470

[131] National Cholesterol Education Program. Detection, evaluation, and treatment of high blood cholesterol in adults (Adult Treatment Panel III): Final report. Bethesda, MD: National Institutes of Health, National Heart, Lung and Blood Institute; 1993. *NIH Publication* No. 02-5215.

[132] Feinman RD, Fine EJ: Thermodynamics and Metabolic Advantage of Weight Loss Diets. *Metabolic Syndrome and Related Disorders* 2003, 1:209-219.

[133] Volek JS, Westman EC: Very-low-carbohydrate weight-loss diets revisited. Cleve *Clin J Med* 2002, 69(11):849, 853, 856-8

[134] Westman EC, Mavropoulos J, Yancy WS, Volek JS: A Review of Low-carbohydrate Ketogenic Diets. *Curr Atheroscler Rep* 2003,5(6):476-483.

[135] Garg A, Bantle JP, Henry RR, Coulston AM, Griver KA, Raatz SK, Brinkley L, Chen YD, Grundy SM, Huet BA: Effects of varying carbohydrate content of diet in patients with non-insulin dependent diabetes mellitus. *Jama* 1994, 271(18):1421-1428.

[136] McLaughlin T, Reaven G, Abbasi F, Lamendola C, Saad M, Waters D, Simon J, Krauss RM: Is there a simple way to identify insulin resistant individuals at increased risk of cardiovascular disease? *Am J Cardiol* 2005, 96(3):399-404.

[137] Volek JS, Sharman MJ, Gomez AL, Scheett TP, Kraemer WJ: An Isoenergetic Very Low-Carbohydrate Diet Is Associated With Improved

Serum High-Density Lipoprotein Cholesterol (HDL-C), Total Cholesterol to HDL-C Ratio, Triacylglycerols, and Postprandial Lipemic Responses Compared to a Low-Fat Diet in Normal Weight, Normolipidemic Women.*J Nutr* 2003, 133(9):2756-2761

[138] Allick G, Bisschop PH, Ackermans MT, Endert E, Meijer AJ, Kuipers F, Sauerwein HP, Romijn JA: A low-carbohydrate/high-fat diet improves glucoregulation in type 2 diabetes mellitus by reducing post-absorptive glycogenolysis. *J Clin Endocrinol Metab2004*, 89(12):6193-6197

[139] Gordon DJ, Probstfield JL, Garrison RJ, Neaton JD, Castelli WP, Knoke JD, Jacobs DRJ, Bangdiwala S, Tyroler HA: High-density lipoprotein cholesterol and cardiovascular disease. *Four prospective American studies. Circulation* 1989;79:8–15. [PubMed: 2642759]

[140] Lamarche B, Tchernof A, Moorjani S, Cantin B, Dagenais GR, Lupien PJ, DespresJP. Small dense low-density lipoprotein particles as a predictor of the risk of ischemic heart disease in men. Perspective results from the Quebec cardiovascular study. *Circulation* 1997;

[141] Hokanson JE, Austin MA: Plasma triglyceride level is a risk *J Cardiovasc Risk.* 1996 Apr;3(2):2139.

[142] Stampfer MJ, Krauss RM, Ma J, Blanche PJ, Holl LG, Sacks FM, Hennekens CH: A prospective study of triglyceride level, low-density lipoprotein particle diameter, and risk of myocardial infarction. *JAMA.* 1996 Sep 18;276(11):882-8

[143] Lau C, Faerch K, Glümer C, Tetens I, Pedersen O, Carstensen B, Jørgensen T, Borch-Johnsen K; Inter99 study. Dietary glycemic index, glycemic load, fiber Diabetes Care. 2005 Jun;28(6):1397-403. *Erratum in: Diabetes Care.* 2005 Sep;28(9):2340-1.

[144] Liese AD, Schulz M, Fang F, Wolever TM, D'Agostino RB Jr, Sparks KC, Mayer-Davis EJ: Dietary glycemic index and glycemic load, carbohydrate and fiber intake, and measures of insulin sensitivity, secretion, and adiposity in the Insulin Resistance Atherosclerosis Study. *Diabetes Care.* 2005 Dec;28(12):2832-8.

[145] Riccardi G, Rivellese AA: Effects of dietary fiber *Diabetes Care.* 1991 Dec;14(12):1115-25. Review

[146] Rivellese AA, Auletta P, Marotta G, Saldalamacchia G, Giacco A, Mastrilli V, Vaccaro O, Riccardi G: Long term metabolic effects of two dietary methods of treating hyperlipidaemia *BMJ.* 1994 Jan 22;308(6923):227-31.

[147] Shah M, Garg A: High-fat *Diabetes Care.* 1996 Oct;19(10):1142-52. Review.

[148] Hung T, Sievenpiper JL, Marchie A, Kendall CW, Jenkins DJ: Fat versus carbohydrate in insulin resistance, obesity, diabetes and cardiovascular disease. *Curr Opin Clin Nutr Metab Care.* 2003 Mar;6(2):165-76

[149] Schwenke DC. Insulin resistance *Curr Opin Lipidol.* 2005 Feb;16(1):55-60. Review.

[150] McAuley KA, Smith KJ, Taylor RW, McLay RT, Williams SM, Mann JI Long-term effects of popular dietary approaches on weight loss_*Int J Obes* (Lond). 2006 Feb;30(2):342-9.

[151] McAuley K, Mann J: Thematic review series: patient-oriented research. Nutritional determinants of insulin resistance. *J Lipid Res.* 2006 Aug;47(8):1668-76. Epub 2006 May 23

[152] Hu FB, Willett WC: Optimal diets for prevention of coronary heart disease. *JAMA*. 2002 Nov 27; 288(20):2569 078

In: Saturated Fats
Editor: Joseph P. Langella
ISBN: 978-1-61209-566-0

Chapter II

Modulation of Dietary Conjugated Linoleic Acid (CLA) Effects by Saturated Fats

Paula Alexandra Lopes, Susana Vargas Martins and José A. Mestre Prates
CIISA, Faculdade de Medicina Veterinária,
Universidade Técnica de Lisboa, Avenida da Universidade Técnica,
Pólo Universitário do Alto da Ajuda, Lisboa, Portugal

1. Introduction

Today's unhealthy diet and lifestyles promote excessive intake of saturated fats (WHO, 2003). High fat diets, possessing inappropriate portions of unsaturated *versus* saturated fatty acids (SFA), may directly increase the prevalence of obesity worldwide (Li et al., 2008). In contrast to SFA, dietary polyunsaturated fatty acids (PUFA) have been recognized as capable of reducing the accumulation of abdominal adipose tissue and lipids in the liver and serum, and also to alleviate hypertension and type 2 diabetes, through the regulation of many enzymes involved in both lipid and glucose metabolisms (Nagao and Yanagita, 2008). One unsaturated fatty acid that has attracted much of the scientific community interest is the conjugated linoleic acid (CLA), a group of minor fatty acids of ruminant origin with potential beneficial effects for long term human health. Interestingly, some studies revealed that CLA is more beneficial in controlling adiposity and blood lipids when supplemented to a diet rich in saturated fat *versus* unsaturated fat (Kloss et al., 2005). In fact, a complex interaction between dietary manipulations (such as, type and levels of fat) and the body fat-reducing action of CLA has been already reported (Hargrave et al., 2004; Akahoshi et al., 2005). However, research performed so far has not adequately compared the biological effects of CLA and the respective molecular mechanisms of action when supplemented to diets containing distinct sources of saturated fat.

In this book chapter, we report our own scientific journey concerning the disclosure of CLA biological effects and the underlying mechanisms of action upon dependence on saturated fat dietary matrices with differential lipid compositions. Two distinct experimental trials were carried out with normal-weight Wistar and obese Zucker rats fed saturated fat based diets supplemented with CLA isomers (*cis*(*c*)9,*trans*(t)11 and *t*10,*c*12, which are the known physiologically active isomers), separately or combined. New (molecular) insights towards CLA action were exploited focusing, in particular, on lipid and glucose metabolisms, inflammatory responses, apoptosis and involved signaling pathways, as well as biomembrane properties associated with cellular trafficking events.

2. Some Biological Effects of Conjugated Linoleic Acid

Increased consumption of highly energetic food with great levels of saturated fats have led to a threefold increase of obesity rates in industrialized societies, contributing to human morbidity and premature mortality (Adams et al., 2006). Facing this worldwide health problem, new compounds that might prevent or even reverse such scenario have attracted scientific attention. As it is difficult to reach nutritional targets for dietary fat consumption, the use of natural compounds with bioactive properties able to produce functional foods, constitutes an interesting strategy to attenuate and prevent obesity incidence. One of the natural compounds most studied in the last decades as a body fat reducing agent is the CLA. It is amazing the burst of scientific investigation depicted by this minor fatty acid, commonly known as CLA, for nearly three decades. CLA conveys potential benefits to long term human health, including cancer prevention, improved immune response, reduced atherosclerotic risk and obesity incidence, among others.

CLA can be a single isomer or the collective term for a mixture of fatty acids containing 18 carbons and at least 2 double bonds separated by a single bond, in either *cis* or *trans* configuration. This nomenclature first arrived from the synthetic origin of these fatty acids from alkali isomerization of linoleic acid (18:2*c*9,*c*12) into *c*9,*t*11 (the rumenic acid) and *t*10,*c*12. On the other hand, CLA is found in beef and dairy products. This natural presence of CLA in ruminant animals involves sequential steps of enzymatic isomerization and reduction of PUFA, desaturation of MUFA, and elongation and β-oxidation of CLA isomers. The main CLA isomer in ruminant fat, with at least 60%, is the *c*9,*t*11, the second most prevalent is usually the *t*7,*c*9, plus many others resulting from the aforementioned enzymatic processes occurring in the rumen bacteria. The content of the *t*10,*c*12 CLA isomer in food is somewhat vestigial (Martins et al., 2007).

In the 90´s, the National Academy of Sciences of the USA recognized CLA as the only fatty acid that unequivocally inhibits carcinogenesis in experimental animal models (National Research Council, 1996). The mechanism of carcinogenesis modulation by CLA is not completely understood, although it may be related to its anti-oxidative properties or to the induction of apoptotic cell death and cell-cycle regulation (Yamasaki et al., 2006). A growing number of experimental studies also evidence beneficial effects in atherosclerosis prevention, improvement of immunological function, and modulation of diabetes and body composition, specially, by reducing body fat and increasing lean fat masses. However, if some reports describe health promising benefits, others reveal lipodistrophy, fatty liver and insulin resistance, after CLA consumption (reviewed by Park and Park, 2010). Moreover, specific physiological effects have been linked to individual CLA isomers. While *c*9,*t*11 and *t*10,*c*12 CLA isomers seem to be equally effective in preventing carcinogenesis, only the *t*10,*c*12 seems to play an important role on lipid metabolism (Pariza et al., 2001). At this point, it is important to acknowledge that dietary supplements have a different CLA isomeric profile compared with food products. The main difference concerns the high percentage of *t*10,*c*12 isomer in these formulas and some studies have recently demonstrated adverse effects of this isomer to human health (Risérus et al., 2002). The commercial synthetic CLA mixtures contain more than 80% of equal amounts of *c*9,*t*11 and *t*10,*c*12 isomers with the premise of reducing human body fat mass and increasing lean fat mass. In line with this, numerous reports have demonstrated that dietary CLA can interfere with body fat composition, in particular the *t*10,*c*12 isomer, by reducing white adipose mass in mice models and cell culture systems, through distinct actions on fat deposition, lipolysis of adipose cells and overall lipid metabolism. Whilst the majority of reports published so far examined the joint effects of *c*9,*t*11 and *t*10,*c*12 isomers as major components, increasing evidence suggest that the physiological effects of CLA are isomer specific as well as tissue related. The diversity of CLA isomer structures reflects marked variations in their biological responses. The chemical structure/activity relationship indicates that a system of conjugated double bonds might be required for its biological function. This hypothesis was already validated *in vitro* for long-chain conjugated fatty acids with 19, 20 and 22 carbons of synthetic origin (Park and Park, 2010). Having this discussion in mind, our first *in vivo* experimental assay attended to this particular hypothesis (see point 4).

3. Nutritional Aspects of Saturated Fats from Vegetable and Animal Origins

In recent years, the level of SFA in human diet has been the subject of special concern among nutritionists. Nutritional guidelines with the target of reducing the incidence of cardiovascular diseases (CVD) have advocated the reduction of total fat, SFA and *trans* fatty acids (TFA) intakes (USDA & HHS, 2005). There have been many reports stating that a high-fat diet, and, particularly the consumption of SFA, impairs insulin sensitivity (Kim et al., 2000; Vessby et al., 2001). Total dietary fat impacts insulin sensitivity in a considerable magnitude, such that high fat diets (>40% fat) worse insulin resistance, whereas low fat diets (typically <20% fat) improve insulin action (Lovejoy et al., 1998). There is strong evidence indicating that dietary fat subtypes have the ability to modulate insulin action directly by changing fatty acid composition of phospholipids. Changes in the lipid profile of cell membranes caused by dietary lipids can affect membrane structure and fluidity, signaling transduction pathways mediated by lipids, and the degree of lipid peroxidation in cell membranes (Hulbert et al., 2005). The variations in membrane fluidity due to PUFA enrichment may produce functional changes of specific integral and membrane-bound proteins, because their lateral mobility, conformation and interaction with other components can be altered (Escrich et al., 2007). For instance, PUFA enrichment has been shown to reduce the number of sodium channels, whereas saturated and *trans* unsaturated fatty acids display an opposite effect (Spector and Burns, 1987). Intake of saturated fats is strongly linked to the development of obesity and insulin resistance, while that of PUFA is not. This is consistent with observations that saturated fats are poorly oxidized for energy and thus readily stored, are weakly mobilized by lipolytic stimuli, impair membrane function, and increase the expression of genes associated with adipocyte proliferation (Storlien et al., 2000). Insulin secretion is differentially affected by various fatty acids *in vitro*. Longer-chain fatty acids and those with greater degree of saturation increase insulin secretion from perfused rat pancreas. In mouse islet cells, TFA potentiate insulin secretion relative to *cis* isomers of the same chain length (Lovejoy et al., 2002). Serum levels of SFA containing a short carbon atom chain, namely lauric (12:0), myristic (14:0), palmitic (16:0) and stearic (18:0) acids, are positively correlated with the homeostasis model insulin resistance index (HOMA-IR) (Kusunoki et al., 2007).

Foodstuffs from vegetable or animal origin convey distinct dietary fat subtypes and, therefore, the nutritional consequences of their intake are discrepant. Concerning saturated fats, palm oil, rich in palmitic acid, is included in several processed instead of partially hydrogenated oils, in order to avoid the TFA increase (Edem, 2002). In fact, crude palm oil is the second largest consumed vegetable oil in the world (Edem, 2002). Palm oil and partially hydrogenated soybean oils, compared with soybean and canola oils, adversely increased serum LDL-cholesterol levels from moderately hyperlipidemic subjects, without significantly affecting glucose and HDL-cholesterol levels (Vega-López et al., 2006). However, other studies failed to demonstrate elevated plasma cholesterol levels following palmitic acid consumption. Clandinin and co-workers (1999) have suggested that other fatty acids influence the cholesterolaemic effect of palmitic acid. When the dietary level of linoleic acid was low, increasing the level of palmitic acid produced a cholesterol-raising effect in plasma. However, when the linoleic acid level was increased to the current dietary recommended level (10% of energy), the effect of increasing palmitic acid had no significant consequence on serum total or LDL-cholesterol levels. Reynoso et al. (2003) reported that high levels of palmitic acid lead to insulin resistance due to changes in the level of phosphorylation of the insulin receptor and insulin receptor substrate-1.

Meat is among the major dietary sources of myristic acid and makes a relevant contribution to palmitic acid intake. Both of these fatty acids are known to develop atherogenic effects. Meat also contributes to the intake of stearic acid, claimed to be highly thrombogenic (Renaud et al., 1979), thus justifying a recommendation to restrict its consumption. Interestingly, some data alert to positive or absent effects of stearic acid. Kelly et al. (2001) showed that diets enriched in this fatty acid did not contribute to an increase in classical cardiovascular risk factors and those related to thrombogenesis. Louheranta et al. (1998) observed that a diet rich in stearic acid did not deteriorate glucose tolerance or insulin action in young healthy females as compared with a diet enriched in oleic acid. Developing public health policies to promote a decrease in meat consumption, in part due to the deleterious effects of SFA in health, neglects the value of these foods as an essential source of some micronutrients (Biesalski, 2005) and their pivotal role as functional foods (Jiménez-Colmenero et al., 2001). Guidelines for reducing levels of dietary SFA have been mainly supported by data from epidemiological studies indicating that high saturated fat intake is associated with concomitant high incidence of CVD. However, recent evidence from clinical studies is not as strong or consistent in pointing out that saturated fat intake is the main cause of CVD (Doyle, 2004). Meat provides unsaturated fatty acids with recognized health benefits, including CLA and *n*-3 PUFA. Regarding other animal lipid sources, edible fats from ruminant animals are the major contributor to saturated fat intake (Givens, 2005) but also contain numerous isomers of oleic and linoleic acids originated from incomplete rumen biohydrogenation of these fatty acids. Some biohydrogenation products are *trans* MUFA that possess contrasting effects on human health (Prates and Bessa, 2008). Vaccenic acid (18:1*t*11) is metabolized into 18:2*c*9,*t*11 CLA isomer (Lock et al., 2004), and for this reason has been considered as beneficial or neutral. Recent data suggest that vaccenic acid intake may convey health benefits beyond those associated with CLA (Field et al., 2009). Either way, the biological properties of TFA from ruminant fat origin are still poorly exploited and are often considered as unhealthy lipids (Willet, 2006). At least one bond (or more) in the *trans* configuration raises the ratio of plasma LDL to HDL cholesterol, regardless of their origin or structure (Brouwer et al., 2010). Epidemiological studies suggest that chronic consumption of *trans* MUFA may alter muscle insulin sensitivity. However, Tardy et al. (2008) concluded that dietary *trans* MUFA of dairy or industrial origin have similar effects and do not impair muscle mitochondrial capacity and insulin sensitivity.

4. Effects of Conjugated Linoleic Acid and Palm Oil on Normal-Weight Wistar Rats

In a feeding trial of eight weeks, Wistar male rats, a normolipidemic and non-obese laboratory rodent model, aging 28 days were allocated to four saturated fat dietary groups based on 5.1% of palm oil and supplemented with both CLA isomers, *c*9,*t*11 and *t*10,*c*12, individually (0.6% w/w) or combined in equal proportions (1.3% w/w). A control group, non-supplemented with CLA, was included for comparative purposes.

4.1. Influence on the Fatty Acid Composition of Hepatic Polar and Neutral Lipids

The rat is a highly valuable model for research on metabolic disorders, as the ones concerning lipid metabolism (Van Zutphen et al., 2001), and the liver, the adequate organ to investigate CLA and other fatty acids profile. In fact, liver plays a central role in lipid metabolism, and possesses a reasonable ratio between polar and neutral lipid fractions (Banni et al., 2001). We looked for the pattern of fatty acids deposition, and, in particular, the detailed profile of CLA isomers incorporation into polar and neutral hepatic lipids by GC and HPLC techniques. In agreement with literature, the deposition of *c*9,*t*11 CLA isomer was much higher than the *t*10,*c*12 isomer, regardless the lipid fraction. Both CLA isomers deposit preferentially into the neutral lipid fraction and do not behave like linoleic acid. For the reason that linoleic acid concentration in tissue is higher than that of CLA, the extension of linoleic acid available may tend to exclude CLA from incorporation into phospholipids, driving it to storage as a component of the neutral lipid fraction, as proposed by Banni et al. (2001). Alternatively, there might be discriminating controls for linoleic acid and CLA uptake into polar lipid classes (Hoffmann et al., 2005), not determined so far. Whilst the major supplemented CLA isomers were preferentially deposited into the neutral lipids, other minor CLA isomers were differentially incorporated into both hepatic lipid fractions. However, the *c*9,*t*11 CLA isomer deposition was higher in the mixture group, relative to the *c*9,*t*11 group. In contrast, the *t*10,*c*12 CLA isomer incorporation was similar for the mixture and *t*10,*c*12 dietary groups. Rat livers from the mixture group had the lowest *n*-6 long chain PUFA but the highest SFA contents in the polar and neutral lipid fractions, respectively. These findings suggest that under these experimental conditions, dietary CLA isomers modulate the fatty acid profile in the liver. This fact may convey beneficial outcomes on the eicosanoid production and, ultimately, benefit the inflammatory response (unpublished data).

4.2. Looking for Morphometrical Features of Adipocyte Cells from Visceral Fat Tissue

Another approach using the same experimental design was carried out by performing a histometrical study, using visceral white adipose tissue from Wistar rats, through the assessment of both number and size of cell adipocytes. We hypothesized that palm oil based diets supplemented with *c*9,*t*11 and *t*10,*c*12 CLA isomers would contribute to alter adipose mass weight and fat cell size and, therefore, change the levels of circulating adipocytokines. A striking finding was that the *c*9,*t*11 isomer, in contrast to the *t*10,*c*12 isomer and the mixture of both isomers, increased adipocyte dimensions in epididymal and retroperitoneal fat depots, through biochemical pathways that remain presently unclear. This effect was followed by an increase of circulating triacylglycerols, while having minor consequences on serum adipocytokines, including leptin and adiponectin contents (Lopes et al., 2008). In view of the fact that CLA mixture did not affect rat´s fat composition and adipocyte size, it is reasonable to speculate that somehow the *t*10,*c*12 isomer, in these particular conditions, counterbalanced the effects of the *c*9,*t*11 isomer. In line with this, our study validated the *t*10,*c*12 isomer as having beneficial effects for reducing cell adipocytes size.

4.3. Disclosing Water and Glycerol Permeabilities in Kidney Membranes

We went further on trying to disclosure the mechanisms of action of CLA. At this time, we search for its influence on cell membrane transport properties, a research line of investigation for which CLA was nearly unrevealed. The effect of dietary CLA isomers on water and glycerol permeabilities was investigated using kidney proximal tubule membranes. Fatty acyl chains are the major components of complex biological membranes, both quantitatively and qualitatively, accounting for more than a half of the mass of most phospholipids and are primarily responsible for the apolar nature of the membrane bilayer. Depending on their chain length and degree of unsaturation, they contribute to fluidity and other physical and chemical properties of the membranes. PUFA are important constituents of membrane phospholipid bilayers, and PUFA composition of the cell membrane influences its physical properties and cellular functions, such as endocytosis, exocytosis, the activities of membrane-associated enzymes and transport. The preparation of isolated brush-border membrane vesicles (BBMV) provides the possibility of studying transport processes independently of other cellular events. Moreover, being more resistant and viable than intact cells, vesicles can be prepared with any chosen internal media, expanding the range of experimental conditions. Using purified kidney BBMV preparations, membrane permeabilities for water and glycerol were assessed by stopped-flow light scattering. The activation energy of water and glycerol transport was calculated. Additionally, the fatty acid profile of membrane vesicles was determined in order to assess CLA isomers incorporation into total membrane lipids. Our main conclusion was that a palm oil-based diet supplemented with a mixture of *c*9,*t*11 and *t*10,*c*12 CLA isomers in Wistar rats, in contrast to its individual isomers, positively affected the glycerol membrane permeability of kidney proximal tubule, possibly by means of an increase in lipid bilayer fluidity. Our results suggest that the *c*9,*t*11 isomer intake promoted a higher incorporation of the *t*10,*c*12 into the membrane, thereby increasing the CLA amount available. In addition to this, the diversity of CLA isomers found in kidney BBMV from CLA mixture may alter fatty acids arrangement and their packing in the bilayer changing membrane fluidity, thus explaining the observed increase in glycerol permeability. An increase in bilayer permeability has also been reported for liposomes containing CLA isomers (Yin et al., 2006). This fact may be of biological relevance because in the kidney, glycerol can be either metabolized *in situ* or converted to glucose in the liver. Conversely, the data indicate that CLA isomers, isolated or jointed, do not affect aquaporin-mediated water transport (Soveral et al., 2009).

5. Effects of Conjugated Linoleic Acid and Vegetable or Animal Saturated Fats on Obese Zucker Rats

In order to search for the mechanisms of CLA action involved in lipid metabolism and lipotoxicity induced by saturated fatty acids (Wei et al., 2007), a second *in vivo* experiment was performed with an obese mutant laboratory animal, the *fa*/*fa* Zucker rat. These animals were fed with a mixture of *c*9,*t*11 and *t*10,*c*12 CLA isomers (1% w/w, alike nutritional supplements) supplemented to saturated fat-based diets of vegetable and animal origins, palm oil and ovine fat, respectively. Contrarily to palm oil, the ovine fat is not only a rich saturated fatty acid source but also have natural conjugated and non-conjugated linoleic isomers and several *trans* MUFA compounds. The *fa*/*fa* Zucker genotype is an obese, leptin receptor-deficient model that exhibits hyperinsulinemia without hyperglycemia (Zucker and Zucker, 1961), non-alcoholic fatty liver disease, and hypertriglyceridemia due to an increased hepatic production of very low density lipoproteins (Bray, 1997). Also, featuring these rats are adipocyte hyperplasia and hypertrophy (Marques et al., 1998). Thus, the Zucker rat is an excellent model for obesity research (Kurtz et al., 1989; Koteish and Diehl, 2001). In a feeding trial of 14 weeks, thirty-two male Zucker rats aging 6 weeks old were fed semi-purified atherogenic diets (Provimi Kliba, SA, Switzerland), with 2% (w/w) of cholesterol and 15% (w/w) of fat with distinct fatty acid composition (Martins et al., 2010a). Rats were divided into four groups with eight animals each, according to the dietary fat: group P - 11.3% palm oil (w/w) plus 3.8% (w/w) sunflower oil; group PCLA - 11.3% (w/w) palm oil plus 2.5% sunflower oil (w/w) plus 1.2% (w/w) CLA; group O - 11.3% (w/w) ovine fat plus 3.8% (w/w) sunflower oil; group OCLA - 11.3% (w/w) ovine fat plus 2.5% (w/w) sunflower oil plus 1.2% (w/w) CLA. The CLA oil contained similar proportions of *c*9,*t*11 and *t*10,*c*12 isomers with a purity level of 80%.

5.1. Assessing Deposition of CLA Isomers and Other Fatty Acids in Hepatic and Muscle Tissues

Initially, our purpose was to investigate the combined effects of CLA and saturated fats from distinct sources, vegetable or animal, on the fatty acid profile from metabolic relevant tissues, as the liver and muscle of obese Zucker rats. Differential deposition of CLA isomers in rat tissues has not been adequately reported yet. In CLA fed rats, the SFA content in the liver increased, whereas the MUFA content decreased. The percentage of PUFA in the liver remained unchanged. In the muscle, CLA did not affect SFA but decreased MUFA and PUFA contents. The TFA were enhanced by CLA supplementation. The estimation of Δ9-index16 and Δ9-index18 suggested that CLA inhibited the Δ9-desaturase activity in the liver, and, specially, when supplemented to the ovine fat diet. The Δ9-desaturation of vaccenic acid is the main pathway of CLA endogenous synthesis (Corl et al., 2001). In the liver, fatty acids from *n*-3 series were increased by CLA when supplemented to ovine fat diet. In the muscle, CLA supplementation diminished *n*-6 long chain fatty acids. In both selected tissues, the *t*10,*c*12 CLA isomer was less incorporated than the *c*9,*t*11 CLA isomer. Nevertheless, the *t*10,*c*12 isomer was twofold higher in the muscle than in the liver. The ingestion of ovine fat based diets resulted in the incorporation of several ruminant isomers.

In a general overview of this study, the liver and muscle fatty acid profiles of rats fed palm oil and ruminant fat were found to enclose interesting interactions between CLA and fat source. The modification of Δ9-desaturase index showed a different influence on Δ9-desaturase activity in both liver and muscle by CLA. Through unclear molecular pathways, CLA also increased the 18:1*t*10 content. There were significant changes in *n*-6 and *n*-3 hepatic and muscle fatty acids when CLA was fed to the animals. The percentage of *c*9,*t*11 isomer was higher in the liver than in the muscle. On the contrary, *t*10,*c*12 isomer was twofold higher in the muscle than in the liver, which reveals different metabolic pathways of CLA isomers in both tissues. These results have shown that CLA supplementation in obese Zucker rats altered the fatty acid profile and most importantly, was dependent upon dietary fat source (unpublished data).

5.2. Inflammatory Response and Related Fatty Acid Composition of Adipose Tissues

CLA has also been reported as having the ability to modulate the inflammatory system in several animal models. Subsequently, the effects of CLA added to saturated fat diets, from vegetable or animal origins, on the serum adipocytokine profile of obese Zucker rats were assessed. In addition, the fatty acid composition of epididymal and retroperitoneal adipose tissues were determined and a principal component analysis (PCA), a statistical discriminant tool, was used to evaluate possible relationships between fatty acids and serum pro- and anti-inflammatory markers. Several CLA effects were observed independently of the saturated fat type present in the diet. CLA fed animals exhibited lower daily feed intake, final body and liver weights, and hepatic lipids content. Surprisingly, total and LDL-cholesterol levels were increased in CLA supplemented groups without affecting serum triacylglycerols. CLA also promoted higher adiponectin and lower plasminogen activator inhibitor-1 (PAI-1) serum levels. Neither CLA nor saturated fat type affected systemic concentrations of ghrelin, C-reactive protein (CRP) and monocyte chemotactic protein (MCP)-1.

The intake of atherogenic diets with 2% of cholesterol along with changes found for cholesterol profile and PAI-1 led us to look for possible alterations in aorta microvasculature from these rats. We had the expertise of a pathologist blinded to group assignment or any biochemical results. The histological examination of different transversal segments of aortas was carried out by immersion in 10% neutral buffered formalin followed by paraffin inclusion. Tissue sections (with 10 μm thick) were cut on a microtome from each of the paraffin-embedded specimens. Sections were then stained with classical hematoxylin and eosin to assess morphology under a light microscope. The histology revealed only small morphological changes affecting smooth muscle cells. Yet, no atheroma plaques were observed for any dietary group. This result is not surprising because rodents are not the suitable model for atherosclerosis studies (Narayanaswamy et al., 2000). Species strain, body size, and average blood pressure are key factors for atherosclerosis appearance and disease progression. These former characteristics make the rabbit, among all laboratory animal models, the ideal species to study human atherogenesis (Narayanaswamy et al., 2000).

In contrast to palm oil based diets, the presence of ovine fat increased insulin resistance and serum levels of leptin, tumour necrosis factor (TNF)-α and interleukin (IL)-1β. Ovine fat diets contain minor amounts of TFA derived from rumen biohydrogenation, *trans* and *cis* isomers of oleic acid and conjugated and non-conjugated *trans* isomers of linoleic acid, which can display additional roles on these effects. Epididymal and retroperitoneal adipose tissues had similar deposition of individual fatty acids. The PCA analysis showed that the *t*10,*c*12 isomer was highly associated with adiponectin and PAI-1 serum levels. Summing up, CLA added to vegetable saturated enriched atherogenic diets, relative to those from animal origin, seemed to improve the serum profile of adipocytokines, specifically the inflammatory markers as well as to alleviate hepatic steatosis, through hepatic lipids reduction in obese Zucker rats due to a more favorable fatty acid composition (Martins et al., 2010a).

5.3. Exploiting Apoptosis in Visceral Fat Tissue and Lipotoxic Liver

In the next study, we hypothesized that reducing weight properties of CLA are due to adipocyte apoptosis and that CLA differentially modulates the apoptotic responses in hepatic lipotoxicity from Zucker rats. To the best of our knowledge, this was the first report on the influence of feeding CLA and saturated fat diets from distinct origins, in cellular apoptosis of visceral adipose tissue and liver, using an obese research model. Firstly, we asked if CLA can promote loss of body fat through adipocyte apoptosis, despite the genetics of the experimental model. The cell-death counting results for retroperitoneal fat revealed no significant differences among dietary groups. This result is consistent with similar fat depots weight. Secondly, facing a lipotoxicity condition, feature of steatotic liver, surely aggravated by atherogenic diets intake, we looked for the influence of CLA supplementation and saturated fats from different sources on the apoptotic molecular responses. All rats presented about 20% of hepatic lipids, which is indicative of an acute lipotoxicity process in the liver. Hepatocyte apoptosis is recognized as an important mechanism contributing to the progression of non-alcoholic fatty liver disease (Feldstein and Gores, 2005). The inhibition of apoptosis represents a valuable therapeutic target to treat human liver diseases (Ribeiro et al., 2004). The apoptotic pathways exploited revealed a differential influence of dietary fatty acids, according to their origin. The higher levels of apoptosis found in palm oil diet with CLA appeared dependent on the extrinsic pathway through up regulation of the cell death receptor Fas. In addition, Fas-mediated hepatocyte apoptosis can lead to up regulation of pro-inflammatory cytokines (Faouzi et al., 2001) and therefore, may contribute to inflammation. Contradicting this hypothesis, the serum pro-inflammatory chemokines levels were kept unchanged in palm oil groups. The ability of apoptotic hepatocytes to synthesize peptide mediators that can recruit inflammatory cells violates one supposed dogma of apoptosis. The basic principle behind the signaling events during apoptotic cell death is that once the point of return is reached, that is, activation of downstream caspases cascade (which comprises the effector caspase-3) a rapid breakdown of the cell is organized (Green, 2000). This happens without eliciting an inflammatory response and avoiding the waste of energy for useless repair processes. In contrast, the ovine fat group with CLA showed less apoptosis than ovine fat group alone, but higher secretion of endoplasmic reticulum (ER)-signaling proteins (ATF-6α, CHOP and BiP) and activation of c-Jun NH_2-terminal kinase (JNK) pathway (p-JNK and p-c-Jun), suggesting an adaptive response of cell survival. These findings reinforce the role of CLA as regulator of apoptosis in the liver. Moreover, fatty acid composition of matrix-diets is a key factor in activation of apoptosis (unpublished data).

5.4. Disclosing Water and Glycerol Permeabilities in Visceral Adipose Tissue

Following on our previous work on biomembranes (Soveral et al., 2009), we also aimed to assess the effects of dietary CLA as a mixture of both isomers, *c*9,*t*11 and *t*10,*c*12, on adipose plasma membrane properties using the same experimental design. The starting point of these *in vitro* experiments comprised the optimization of a procedure in order to obtain isolated plasma membranes from rat's visceral adipose tissue that, being osmotically responsive, are suitable for transport experiments. Western blots from the final adipose membrane preparations were incubated with antibodies reactive to GLUT4, caveolin and aquaporin 7 (AQP7). An intense band of GLUT4, the insulin-stimulated glucose transporter (Kandror et al., 1995), was depicted in the membrane vesicles. Likewise, the band for caveolin, a constitutive protein of plasma membranes of mature adipocytes (Kandror et al., 1995), was stronger in the final vesicle preparations. Glycerol is a key substrate involved in lipogenesis and lipolysis of adipose tissue. Glycerol movements of release and uptake from the adipocyte are mediated by AQP7, a glycerol-channel that regulates glycerol accumulation

(Kishida et al., 2000; Hibuse et al., 2006) and acts like a pressure valve letting some of the glycerol enter the bloodstream (Frühbeck, 2005). If AQP7 is missing or dysfunctional, glycerol accumulates in fat cells, driving to additional synthesis of fat. In our study, AQP7 was also detected in the adipose membranes, despite in a fainter band than caveolin. Altogether, these results assured purified and homogeneous adipose plasma membrane preparations. These membrane preparations were hereafter analyzed regarding their lipid composition by GC and HPLC techniques, permeability to water and glycerol by stopped-flow light scatter and fluidity by fluorescence anisotropy. CLA supplementation in obese Zucker rats through its incorporation in membrane phospholipids had an effect on the fluidity of adipose membranes by increasing membrane rigidity both at the centre of the bilayer (hydrocarbon core) and at a shallow position, closer to the water/lipid interface. Despite the same concentration of both CLA isomers in the oil mixture, the *c*9,*t*11, which is the natural occurring isomer in ruminant fat, was detected in a larger amount than the *t*10,*c*12 in all dietary groups, independently of dietary fat. This result *per se* might suggest that *c*9,*t*11 isomer could be accountable for the changes in membrane properties reported here. However, given that changes in permeability and fluidity were only observed for CLA supplemented groups and did not depend on the fat origin, the *c*9,*t*11 CLA isomer can be discarded because its concentration was much higher in ovine than in vegetable fat. Therefore, the *t*10,*c*12 CLA isomer might be suggested as the most probable membrane component affecting adipose plasma membrane fluidity and permeability. Interestingly, this CLA isomer is the one with suggested body fat lowering properties in several animal studies (Park and Pariza, 2007). This effect could be explained by the geometric structure of the *t*10,*c*12 CLA isomer, where a considerable bending in the fatty acid chain induces a decrease in the average chain length that, similarly to cholesterol, may affect lipid packing within cellular membranes. This isomer, particularly its conformation, may promote the formation of other type of raft-like structure, different from the cholesterol and sphingolipids rich microdomains, contributing to the experimentally observed decrease on the average membrane fluidity and permeability of the system. Consequently, our results did not point out to membrane fluidification as a possible mechanism, supporting CLA as a body fat reducing agent in obese fat consumers (Martins et al., 2010b).

6. Overview of the Experiments in Normal-Weight Wistar *versus* Obese Zucker Rats

Much scientific evidence is available concerning the distinct biological activities of *c*9,*t*11 and *t*10,*c*12 CLA isomers. To exemplify this statement, in human and mouse studies, the *t*10,*c*12 was the isomer responsible for the reduction of body fat weight (for review, see Park and Pariza, 2007), but depicted a pro-diabetic effect in both models (Risérus et al., 2002; Roche et al., 2002). In order to elucidate the divergent aspects promoted by these two CLA isomers, a reference strain with normal growth development and no predisposition to any pathology was selected – the Wistar rat. This experimental model aimed to mimic a healthy normal weight human subject that consumes CLA supplements on a regular basis.

The dietary supplementation of CLA isomers, individually or in combination to Wistar rats did not result in different growth performances, unaffecting several organs weight, including the main adipose depots. The relative leanness of the Wistar model may be held responsible for the lack of significant effects of dietary CLA on rat's body weight, as reported previously by Azain et al. (2000).

The serum lipid profile and adipocytokines were unaffected by CLA, except for systemic triacylglycerols. Notwithstanding, the morphology of adipocytes was changed by *c*9,*t*11 isomer, in both epididymal and retroperitoneal fat depots. To the best of our knowledge, it was reported for the first time an increase of adipocytes area due to the *c*9,*t*11 isomer (Lopes et al., 2008). It is important to highlight that the CLA mixture did not affect body fat composition or adipocyte morphology, suggesting a mechanism of compensation between *c*9,*t*11 and *t*10,*c*12 CLA isomers effects. Remaining to elucidate are the mechanisms involved, although the parallel increase of serum triacylglycerols suggests consequences at the lipid metabolism level, such as an increase in lipogenesis rate. In the same trial, other research line was further followed to exploit the increase in triacylglycerols synthesis, in particular in what glycerol metabolism concerns. The effect of dietary CLA on the permeability of kidney proximal tubules membranes to glycerol was an original study that resulted in novel important findings (Soveral et al., 2009). CLA mixture promoted an increase in glycerol permeability via the lipid bilayer and not by protein-mediated transport. This result may be attributed to an increase in lipid bilayer fluidity through the incorporation of CLA isomers in the membrane re-arrangement. The lack of differential responses in protein-mediated transport suggests that CLA had no influence on the expression and/or activity of aquaglyceroporin channels, namely AQP7. Kidney is an important organ concerning the reabsorption pathway of glycerol re-directing it into the liver. Once in the liver, glycerol can integrate lipogenesis process or can be converted to glucose. Still, this potential pathway for triacylglycerols increase by the CLA mixture is not consensual with the effective enhancement of triacylglycerols only by the *c*9,*t*11 isomer.

Inconsistent anti-adipogenic effects of CLA have been found in humans and also, in distinct animal models. Up to date, no human trials have tried to disclosure the efficacy of CLA supplementation on body fat mass loss in genetic predisposed obese subjects. Additionally to the environmental factors, some studies have indicated that genetic factors account for a substantial portion of human adiposity (Barness et al., 2007). Among many, human mutations in the leptin gene (Montague et al., 1997) and leptin receptor (Clement et al., 1998) are well characterized, resulting in a severe obesity phenotype, which indicates that leptin is critical to body weight regulation. The *fa*/*fa* Zucker rat is a strain of laboratory animal descendent from the wild rodent model *Rattus norvegicus* (Zucker and Zucker, 1961), used often to study genetic obesity problems. This animal model develops morbid obesity, through the *fa*/*fa* mutation in the extracellular domain of the leptin receptor, which inhibits completely leptin action. This leads to an appetite increase promoting severe insulin resistance (Bray, 1977). According to the same author, obese Zucker rats do not have pancreatic β-cells defect, and do not progress to insulin deficiency similarly to human type 2 diabetes. We designed an *in vivo* experiment with obese Zucker rats fed a hyperlipidemic (high-fat) and atherogenic (high cholesterol) diet. This trial aimed to mimic an obese human sub-population with severe metabolic problems that consumes CLA supplements. However, CLA supplementation accounts for a small part of the total fatty acids ingested through the diet. So, the original hypothesis was that CLA display different actions when supplemented to diets with vegetable or animal saturated fats. Most of the saturated fat intake is attributed to edible fats from ruminant origin (Givens, 2005). Both palm oil and ovine fat display high saturated fat content but different fatty acid composition. Briefly, palm oil is rich in palmitic acid while ovine fat possesses more stearic acid and TFA due to the rumen metabolism.

The results herein presented, revealed interesting effects of CLA and dietary fat source, separately, but also effects denoting an interaction between CLA and fat source. After the 14th week of experimental trial, Zucker rats exhibited massive obesity which was prevalent in the abdominal region. Despite showing great amounts of subcutaneous fat, only epididymal and retroperitoneal adipose depots (visceral fats) were in fact, weighted and studied due to methodological constraints related to isolation difficulties in subcutaneous fat extraction. The adipose tissue located in the visceral compartment, in contrast to the subcutaneous fat, contributes more actively to a pro-inflammatory state and loss of insulin sensitivity in obese subjects (Dolinkóva et al., 2008). In the present work, the fatty acid profile of both epididymal and retroperitoneal fat depots were quite similar. The adipocytokines quantification was performed in serum and not locally, but a differential adipocytokines secretion from different fat depots, and even from the liver, could be expected. Due to the leptin receptor mutation, it was not surprising the high levels of leptin found in the serum of *fa*/*fa* Zucker rats. The leptin action is based on the signaling towards the hypothalamus. In there, a cascade of events is initiated, comprising the simultaneous decrease in appetite and insulin secretion with subsequent increase of metabolic rate (Savage and O'Rahilly, 2002). Naturally, one major consequence of leptin deficiency action is overfeeding (Tritos and Mantzoros, 1997). Beck et al. (2003) described that lean and obese adult Zucker rats feed daily 18 g and 23 g, respectively. Many more physiologic discrepancies were reported between lean (*fa*/+) and obese Zucker (*fa*/*fa*) rats. For example, Oana and co-workers (2005) showed that levels of insulin, adiponectin, triacylglycerols and free fatty acids in the plasma of obese Zucker rats were higher than that of lean rats but glucose levels were not significantly different. Nevertheless, the evaluation of the genetic influence on CLA action was not a goal in this study.

There were positive and negative metabolic effects of CLA in the Zucker rat trial. In fact, CLA increased adiponectin and decreased PAI-1 levels in the serum, as well as, alleviated hepatic steatosis by means of hepatic lipids reduction. In contrast, the serum lipid profile was disrupted by increased levels of total and LDL cholesterol. The presence of two distinct saturated fats, of vegetable and animal origins, in the diet also influenced several important adipocytokines, regardless CLA supplementation. Overall, the clinical state of the animals was aggravated by the ingestion of animal fat, regarding insulin resistance and inflammation. Ovine fat encloses a ruminant fatty acid profile highly consumed in Western societies and these results reinforce the non-beneficial effects of this type of fat.

An overview of the main results promoted by CLA supplementation under the influence of distinct saturated fat based diets is presented in Table 1. Findings were classified into deleterious, neutral and beneficial effects, according to the two experimental trials where they insert to. Nevertheless, some caution is required when reading this summary table. The two experimental trials had completely independent designs thus disabling their direct comparison.

Table 1. Overall findings of CLA and saturated fat based diets effects in both animal models.

Main effects in normal-weight Wistar rat experiment		
Deleterious	**Neutral**	**Beneficial**
↑ serum triacylglycerols by *c*9,*t*11	= daily intake	↑ permeability of kidney membranes to glycerol by CLA mixture
↑ adipocyte size by *c*9,*t*11	= final body weight	
	= organs weight	
	= adipose depots mass	
	= serum leptin and adiponectin concentrations	
	= permeability of kidney membranes to water by *c*9,*t*11, *t*10,*c*12 and CLA mixture	
Main effects in obese Zucker rat experiment		
Deleterious	**Neutral**	**Beneficial**
↑ serum total and LDL cholesterol levels by CLA mixture	= adipose depots mass	↓ daily intake by CLA mixture
↑ serum TNF-α and IL-1β levels by ovine fat	= adipocytes apoptosis	↓ final body weight by CLA mixture
↑ serum insulin resistance index by ovine fat	= serum ghrelin, CRP and MCP-1 concentrations	↓ liver weight by CLA mixture
↑ serum leptin concentrations by ovine fat		↓ liver lipids content by CLA mixture
↓ permeability and fluidity of adipose membranes by CLA mixture		↓ serum PAI-1 concentrations by CLA mixture
↑ membrane rigidity by *t*10,*c*12 incorporation in adipose membranes		↑ serum adiponectin concentrations by CLA mixture
		↓ hepatocyte apoptosis by CLA in ovine fat diet

In the Wistar rat trial, the few differences found were mainly due to *c*9,*t*11 isomer and CLA mixture. In opposition, the obese Zucker rats were affected by CLA supplementation but some effects were dependent on the saturated fat type available in the diet. It is important to underline that both *in vivo* studies did not provide any evidence of CLA´s anti-adipogenic action. Therefore, the present chapter questions the usefulness of CLA isomers to prevent or reduce obesity in Western societies, whereas the intake of animal saturated fats is prevalent.

7. Conclusions and Possible Extrapolation to Humans

The human health effects of CLA, including its potential to reduce body fat mass, became the subject of widespread research for nearly three decades. Still, its efficiency remains controversial (Park and Pariza, 2007). Depending on the country, the estimation of CLA daily intake varies from 15 to 1000 mg (Martins et al., 2007), thus representing a large range of consumption values. In 2003, Parrish et al. suggested that a consumption of 0.8-3.0 g CLA *per* day might provide significant health benefits, a dosage that in most cases is only achievable through diet supplementation. In both experimental trials, 1% of supplemented CLA was meant to mimic nutritional supplement use, besides being the most commonly dosage applied in animal research.

As any model used for human health applications, laboratory animal science offers important limitations when facing extrapolation to humans. Effects in rodents are not the same as in humans, perhaps due to different metabolic rate. In comparison, CLA dietary supplements marketed for weight loss represent an intake of 2-4 g *per* day (Pharma Nord, 2009). Most of the CLA products commercialized as dietary supplements contain 60-90% of CLA, predominantly *c*9,*t*11 and *t*10,*c*12 isomers in the form of either free fatty acids or triacylglycerols. The wide range of CLA dosages and compositions in supplements introduces inconsistencies and adds more difficulties in the interpretation and extrapolation of CLA biological effects to humans. According to a recent review, approximately 50% of the studies reported losses of fat after CLA intake in normal, overweight, and obese subjects (Plourde et al., 2008). A meta-analysis of human studies conducted by Whigham and colleagues (2007), in which CLA was provided as a dietary supplement to test its efficacy in reducing fat mass, concluded that given at a dose of 3.2 g/d, CLA produces a modest loss in body fat in humans. This value is close to the range previously suggested by Parrish et al. (2003). Still, it becomes hard to draw definitive conclusions as most of the studies done so far have differed with respect to isomer, dose and study duration. Not to mention several human studies reporting no statistically significant effects of CLA on fat mass. The explanation for some of these findings relies in part on the absence of statistical power because the treatment duration was too short, there were few subjects, or both (Whigham et al., 2007). New and stronger evidences are needed to definitely establish CLA as an anti-adipogenic agent and therefore, as a natural supplement devoid of harmful effects for fat loss in humans. In fact, some studies in humans revealed adverse CLA effects. Just to name a few, CLA may actually decrease HDL-cholesterol levels and increase markers of inflammation in short-term (Steck et al., 2007). Indeed, CLA has been shown to slightly increase biomarkers of inflammatory disease, including CRP, white blood cell counts, and blood and urinary isoprostanes, a standard indicator of lipid peroxidation (reviewed by Whigham et al., 2007). In obese men with the metabolic syndrome, CLA worsened the insulin resistance (Risérus et al., 2002; Risérus et al., 2004). Following these findings, Stachowska and Chlubek (2009) consider the existence of risk groups for CLA consumption, including in particular, diabetes mellitus, intolerance to glucose and coronary thrombosis. The pursuing of CLA biological effects according to different health status is, at this point, absolutely mandatory. CLA supplements are available for a general population that possesses distinct genetic backgrounds and also, that is exposed to a great variety of environmental factors (not accounted in this book chapter) that could exert some influence in the biological activity of CLA isomers.

Interestingly, in some European countries, CLA enriched food products are already available in the supermarket, covering from milk and yoghurts to margarine. The foodstuff containing CLA is classified as a novel food according to the Novel Food Regulation (N. 258/97/EC), as it is "a food or food ingredient consisting of or isolated from plants and food ingredients isolated from animals" and was not obtained by "traditional propagating or breeding practices and having a history of safe food". In 2008, CLA was recognized by the Food and Drug Administration (FDA) as Generally Regarded as Safe (GRAS). Owing to this status, CLA can now be added in the USA to flavored milks, soya milks, milk based meal replacements, yoghurts, fruit juices and so on.

Closing up this book chapter, beneficial and adverse effects of CLA supplemented to saturated fat based diets, were obtained in both experimental trials herein briefly described. CLA supplements should be carefully used given that our own results confirm the modulation of CLA action depending on the diet lipid composition. These findings demonstrate the ability of different saturated fat dietary matrices to trigger discrepant effects in several biochemical and metabolic pathways. The ability of fatty acids to modulate inflammation is particularly important as an increased number of diseases related to obesity have been shown to have an underlying inflammatory process. Nevertheless, the use of CLA, either as a nutrient or as a drug, needs to ensure that it takes into consideration the exclusion of risk factors associated with other diseases.

Acknowledgments

The experimental work quoted in this chapter was supported by the FCT grants POCTI/CVT/2002/44750, PTDC/CVT/2006/66114 and PTDC/CVT/2008/ 99210.

References

Adams KF, Schatzkin A, Harris TB, Kipnis V, Mouw T, Ballard-Barbash R, Hollenbeck A, Leitzmann MF. Overweight, obesity, and mortality in a large prospective cohort of persons 50 to 71 years old. *N Engl J Med* 2006; 355: 763-778.

Akahoshi A, Koba K, Enmoto R, Nishimura K, Honda Y, Minami M, Yamamoto K, Iwata T, Yamauchi Y, Tsutsumi K, Sugano M. Combined effects of dietary protein type and fat level on the body fat–reducing activity of conjugated linoleic acid (CLA) in rats. *Bios Biot Biochem* 2005; 69: 2409-2415.

Azain MJ, Hausman DB, Sisk MB, Flatt WP, Jewell DE. Dietary conjugated linoleic acid reduces rat adipose tissue cell size rather than cell number. *J Nutr* 2000; 130: 1548-1554.

Banni S, Carta G, Angioni E, Murru E, Scanu P, Melis MP; Bauman DE, Fisher SM, Ip C. Distribution of conjugated linoleic acid and metabolites in different lipid fractions in the rat liver. *J Lipid Res* 2001; 42: 1056-1061.

Bray GA. The Zucker-fatty rat: a review. *Fed Proc* 1997; 36: 148-153.

Brouwer IA, Wanders AJ, Katan MB. Effect of animal and industrial *trans* fatty acids on HDL and LDL cholesterol levels in humans – a quantitative review. *PLoS ONE* 2010; 5: e9434.

Clandinin MT, Cook SL, Konrad SD, Goh YK, French MA. The effect of palmitic acid on lipoprotein cholesterol levels and endogenous cholesterol synthesis in hyperlipidemic subjects. *Lipids* 1999; 34: S121-S124.

Corl BA, Baumgard LH, Dwyer DA, Griinari JM, Phillips BS, Bauman DE. The role of delta9-desaturase in the production of *cis*-9,*trans*-11 CLA. *J Nutr Biochem* 2001; 12: 622-630.

Dolinkóva M, Dóstlová I, Lacinová Z, Michalský ZD, Haluzíková D, Mráz M, Kasalický M, Haluzík M. The endocrine profile of subcutaneous and visceral adipose tissue of obese patients. *Mol Cell Endocrinol* 2008; 291: 63-70.

Edem DO. Palm oil: biochemical, physiological, nutritional, hematological, and toxicological aspects: a review. *Plant Food Hum Nutr* 2002; 57: 319-341.

Escrich E, Moral R, Grau L, Costa I, Solanas M. Molecular mechanisms of the effects of olive oil and other dietary lipids on cancer. *Mol Nutr Food Res* 2007; 51: 1279-1292.

Faouzi S, Burckhardt BE, Hanson JC, Campo CB, Schrum LW, Rippe RA, Maher JJ. Anti-Fas induces hepatic chemokines and promotes inflammation by an NF-KB independent, caspase-3 dependent pathway. *J Biol Chem* 2001; 276: 49077-49082.

Feldstein AE, Gores GJ. Apoptosis in alcoholic and nonalcoholic steatohepatitis. *Frontiers in Bioscience* 2005; 10: 3093-3099.

Field CJ, Blewett HH, Proctor S, Vine D. Human health benefits of vaccenic acid. *Appl Physiol Nutr Metabol* 2009; 34: 979-991.

Frühbeck G. Aquaporin enters the picture. *Nature* 2005; 438: 436-437.

Green DR. Apoptotic pathways: paper wraps stone blunt scissors. *Cell* 2000; 102: 1-4.

Hargrave KM, Meyer BJ, Li C, Azain MJ, Baile CA, Miner JL. Influence of dietary conjugated linoleic acid and fat source on body fat and apoptosis in mice. *Obes Res* 2004; 12: 1435-1444.

Hulbert A, Turner N, Storlien L. Dietary fats and membrane function: implications for metabolism and disease. *Biol Rev Camb Philos* 2005; 80: 155-169.

Hibuse T, Maeda N, Nagasawa A, Funahashi T. Aquaporins and glycerol metabolism. *Biochim Biophys Acta* 2006; 1758: 1004-1011.

Kandror KV, Stephens JM, Pilch PF. Expression and compartmentalization of caveolin in adipose cells: coordinate regulation with and structural segregation from GLUT4. *J Cell Biol* 1995; 129: 999-1006.

Kelly FD, Sinclair AJ, Mann NJ, Turner AH., Abedin L, Li D. A stearic acid-rich diet improves thrombogenic and atherogenic risk factor profiles in healthy males. *Eur J Clin Nutr* 2001; 55: 88-96.

Kim JY, Nolte LA, Hansen PA, Han DH, Ferguson K, Thompson PA, Holloszy JO. High-fat diet-induced muscle insulin resistance: relationship to visceral fat mass. *Am J Physiol Regul Integr Comp Physiol* 2000; 279: R2057-R2065.

Kishida K, Kuriyama H, Funahashi T, Shimomura I, Kihara S, Ouchi N, Nishida M, Nishizawa H, Matsuda M, Takahashi M, Hotta K, Nakamura T, Yamashita S, Tochino Y, Matsuzawa Y. Aquaporin adipose, a putative glycerol channel in adipocytes. *J Biol Chem* 2000; 275: 20896-20902.

Koteish A, Diehl AM. Animal models of steatosis. *Sem Liver Dis* 2001; 21: 89-104.

Kurtz TW, Morris RC, Pershadsingh HA. The Zucker fatty rat as a genetic model of obesity and hypertension. *Hypertension* 1989; 13: 896-901.

Kusunoki M, Tsutsumi K, Nakayama M, Kurokawa T, Nakamura T, Ogawa H., Fukuzawa Y, Morishita M, Koide T, Miyata T. Relation between serum concentrations of saturated fatty acids and unsaturated fatty acids and the homeostasis model insulin resistance index in Japanese patients with type 2 diabetes mellitus. *J Med Invest* 2007; 54: 243-247.

Lopes, PA, Martins, SV, Pinho, MS, Alfaia, CM, Fontes, CM, Rodrigues, PO, Morais, GS, Castro MF, Pinto R, Prates, JAM. Diet supplementation with the *cis*-9, *trans*-11 conjugated linoleic acid isomer affects the size of adipocytes in Wistar rats. *Nutr Res* 2008; 28: 480-486.

Louheranta AM, Turpeinen AK, Schwab US, Vidgren HM, Parviainen MT, Uusitupa MI. A high-stearic acid diet does not impair glucose tolerance and insulin sensitivity in healthy women. *Metabolism* 1998; 47: 529-534.

Lovejoy JC, Smith SR, Champagne CM, Most MM, Lefevre M, DeLany JP, Denkins YM, Rood JC, Veldhuis J, Bray GA. Effects of diets enriched in saturated (palmitic), monounsaturated (oleic), or *trans* (elaidic) fatty acids on insulin sensitivity and substrate oxidation in healthy adults. *Diabetes Care* 2002; 25: 1283-1288.

Lovejoy JC, Windhauser MM, Rood JC, de la Bretonne JA. Effect of a controlled high-fat versus low-fat diet on insulin sensitivity and leptin levels in African-American and Caucasian women. *Metabolism* 1998; 47: 1520-1524.

Marques BG, Hausman DB, Martin RJ. Association of fat cell size and paracrine growth factors in development of hyperplastic obesity. *Am J Physiol* 1998; 275: 1898-1908.

Martins AP, Lopes PA, Martins SV, Madeira A, Santos NC, Prates JAM, Moura TF, Soveral G. Conjugated linoleic acid reduces permeability and fluidity of adipose plasma membranes from obese Zucker rats. *Biochem Biophys Res Commun* 2010b; 398: 199-204.

Martins SV, Lopes PA, Alfaia CM, Ribeiro VS, Guerreiro TV, Fontes CM, Castro MF, Soveral G, Prates JA. Contents of conjugated linoleic acid isomers in ruminant-derived foods and estimation of their contribution to daily intake in Portugal. *Br J Nutr* 2007; 1-8.

Martins SV, Lopes PA, Alfaia CM, Rodrigues PO, Alves SP, Pinto RM, Castro MF, Bessa RJ, Prates JA. Serum adipokine profile and fatty acid composition of adipose tissues are affected by conjugated linoleic acid and saturated fat diets in obese Zucker rats. *Br J Nutr* 2010a; 103: 869-878.

Narayanaswamy M, Wright KC, Kandarpa K. Animal models for atherosclerosis, restenosis, and endovascular graft research. *J Vasc Interv Radiol* 2000; 11: 5-17.

National Research Council. *Carcinogens and anticarcinogens in the human diet.* National Academy Press, Washington DC, 1996.

Pariza MW, Park Y, Cook ME. The biologically active isomers of conjugated linoleic acid. *Progress Lipid Res* 2001; 40: 283-298.

Park Y, Pariza MW. Mechanisms of body fat modulation by conjugated linoleic acid (CLA*). Food Res Int* 2007; 40: 311-323.

Park Y, Park Y. Conjugated nonadecadienoic acid is more potent than conjugated linoleic acid on body fat reduction. *J Nutr Biochem* 2010; 21: 764-773.

Parrish FCJ, Wiegand BR, Beitz DC, Ahn DU, Du M, Trenkle AH. Use of dietary CLA to improve composition and quality of animal-derived foods. In: Sébédio JL, Christie WW, Adlof R (Eds.), *Advances in Conjugated Linoleic Acid Research,* AOCS Press, Champaign, IL, 2003, pp. 189-217.

Pharma Nord. *Informative label on the product "Bioactive CLA Forte" capsules.* ApS Denmark, 2009.

Plourde M, Jew S, Cunnane SC, Jones PJH. Conjugated linoleic acids: why the discrepancy between animal and human studies*? Nutr Rev* 2008; 66: 415-421.

Prates JAM, Bessa RJB. *Trans* and *n*-3 fatty acids. In: Nollet LML, Toldrá F (Eds.), *Handbook of Muscle Food Analysis*, CRC Press, New York, 2008, pp. 399-417.

Reynoso R, Salgado L, Calderón V. High levels of palmitic acid lead to insulin resistance due to changes in the level of phosphorylation of the insulin receptor and insulin receptor substrate-1. *Mol Cell Biochem* 2003; 246: 155-162.

Ribeiro PS, Cortez-Pinto H, Solá S, Castro RE, Ramalho RM, Baptista A, Moura MC, Camilo ME, Rodrigues CM. Hepatocyte apoptosis, expression of death receptors, and activation of NF-KappaB in the liver of nonalcoholic and alcoholic steatohepatitis patients. *Am J Gastroenterol* 2004; 99: 1708-1717.

Risérus U. *Trans* fatty acids, unsulin sensitivity and type 2 diabetes. *Scand J Food Nutr* 2006; 50: 161-165.

Risérus U, Basu S, Jovinge S, Fredrikson GN, Arnlov J, Vessby B. Supplementation with conjugated linoleic acid causes isomer-dependent oxidative stress and elevated C-reactive protein: a potential link to fatty acid-induced insulin resistance. *Circulation* 2002; 106: 1925-1929.

Risérus U, Smedman A, Basu S, Vessby B. Metabolites effects of conjugated isomers of linoleic acid in humans: the Swedish experience. *Am J Clin Nutr* 2004; 79: 1146S-1148S.

Roche HM, Noone E, Sewter C, Bennett SM, Savage D, Gibney MJ, O'Rahilly S, Vidal-Puig AJ. Isomer-dependent metabolic effects of conjugated linoleic acid – insights from molecular markers sterol regulatory element-binding protein-1c and LXRα. *Diabetes* 2002; 51: 2037-2044.

Savage DB, O'Rahilly S. Leptin: a novel therapeutic role in lipodystrophy. *J Clin Invest* 2002; 109: 1285-1286.

Spector A, Burns C. Biological and Therapeutic Potential of Membrane Lipid Modification in Tumors. *Cancer Res* 1987; 47: 4529-4537.

Soveral G, Martins AP, Martins SV, Lopes PA, Alfaia CM, Prates JAM, Moura, TF. Effect of dietary conjugated linoleic acid isomers on water and glycerol permeability of kidney membranes. *Biochem Biophys Res Commun* 2009; 383: 108-112.

Stachowska E, Chlubek D. Conjugated linoleic acids – a safe dietary supplement. *J Chinese Clin Med* 2009; 4: 181-185.

Steck SE, Chalechi AM, Miller P, Conway J, Austin GL, Hardin JW, Albright CD, Thuillier P. Conjugated linoleic acid supplementation for twelve weeks increases lean body mass in obese humans. *J Nutr* 2007; 137: 1188-1193.

Storlien LH, Higgins JA, Thomas TC, Brown MA, Wang HG, Huang XF, Else PL. Diet composition and insulin action in animal models. *Br J Nutr* 2000; 83: S85-S90.

Tardy A, Giraudet C, Rousset P, Rigaudière J, Laillet B, Chalancon S, Salles J, Loreau O, Chardigny J, Morio B. Effects of *trans* MUFA from dairy and industrial sources on muscle mitochondrial function and insulin sensitivity. *J Lipid Res* 2008; 49: 1445-1455.

Tritos NA, Mantzoros CS. Leptin: its role in obesity and beyond. *Diabetologia* 1997; 40: 1371-1379.

Van Zutphen LFM, Baumans V, Beyen AC. *Principles of Laboratory Animal Science.* Revised ed, Elsevier, Amsterdam, 2001.

Vega-López S, Ausman LM, Jalbert SM, Erkkilä AT, Lichtenstein AH. Palm and partially hydrogenated soybean oils adversely alter lipoprotein profiles compared with soybean and canola oils in moderately hyperlipidemic subjects. *Am J Clin Nutr* 2006; 84: 54-62.

Vessby B, Uusitupa M, Hermansen K, Riccardi G, Rivellese AA, Tapsell LC, Nälsén C, Berglund L, Louheranta A, Rasmussen BM, Calvert GD, Maffetone A, Pedersen E, Gustafsson IB, Storlien LH. Substituting dietary saturated for monounsaturated fat impairs insulin sensitivity in healthy men and women: The KANWU Study. *Diabetologia* 2001; 44: 312-319.

Wei Y, Wang D, Pagliassotti MJ. Saturated fatty acid-mediated endoplasmic reticulum stress and apoptosis are augmented by *trans*-10, *cis*-12-conjugated linoleic acid in liver cells. *Mol Cell Biochem* 2007; 303: 105-113.

Whigham LD, Watras AC, Schoeller DA. Efficacy of conjugated linoleic acid for reducing fat mass: a meta-analysis in humans. *Am J Clin Nutr* 2007; 85: 1203-1211.

Yamasaki M, Miyazaki HC, Yamada K. Growth inhibition and apoptotic cell death of cancer cells induced by conjugated linoleic acid. In: Yurawecz MP, Kramer JK, Gudmundsen O, Pariza MW, Banni S (Eds.), *Advances in Conjugated Linoleic Acid Research,* AOCS Press, Champaign, IL, 2006, pp. 141-152.

Yin JJ, Kramer JK, Yurawecz MP, Eynard AR, Mossoba MM, Yu L. Effects of conjugated linoleic acid (CLA) isomers on oxygen diffusion-concentration products in liposomes and phospholipid solutions. *J Agric Food Chem* 2006; 54: 7287-7293.

Zucker LM, Zucker TF. Fatty, a new mutation in the rat. *J Hered* 1961; 52: 275-278.

In: Saturated Fats
Editor: Joseph P. Langella

ISBN: 978-1-61209-566-0

Chapter III

Effects of Short- and Long-Term Saturated Fat-Enriched Diet on the Ceramide and Neutral Lipids Accumulation in the Insulin Responsive Tissues of Rats

Nataliya A. Babenko[*]
Department of Physiology of Ontogenesis, Institute of Biology, Kharkov Karazin National University, Kharkov, Ukraine

Abstract

High-fat diet is a significant risk factor for the development of insulin resistance and Alzheimer`s disease (AD). The link between dietary saturated fat-induced insulin resistance and neurodegeneration and AD has been determined. However, the precise mechanisms underlying these dietary influences are not completely understood. We investigated effects of dietary saturated fats and palmitic acid on ceramide and neutral lipids accumulation in the insulin responsive tissues and insulin-induced glucose uptake. 2-week

* Department of Physiology of Ontogenesis, Institute of Biology, Kharkov Karazin National University, 4 Svobody pl., 61077 Kharkov, Ukraine, E-mail: babenko@univer.kharkov.ua, Phone: +30 57 7075665, +30 57 7062148, Fax: +30 57 3352923.

feeding of rats with saturated fat increased sphingomyelin (SM) and decreased phosphatidylcholine levels and did not change the ceramide and neutral lipids and free fatty acids contents in the liver. The short-term diet also increased the palmitic and oleic acids and decreased the arachidonic acid contents. The dietary saturated fat, chronically added to the food, was found to increase free fatty acids as well as fatty acid metabolites (ceramide and neutral lipids) contents in such insulin key target tissues as liver, muscle and brain. Similar results were obtained in the experiments with palmitate- or C2-ceramide-treated cell and tissue cultures. Treatment of the isolated liver hepatocytes, diaphragm strips and cortex slices of adult rats which had a free access to a standard chow diet with palmitic acid increased the ceramide and neutral lipids contents in the cells and tissues and decreased the insulin-induced glucose uptake by the cells. The palmitic acid effect on insulin-stimulated glucose transport in the tissues studied could be mimicked by the addition of short chain (C2) ceramide to the culture medium. Taken together, our results suggest that long-term exposure of adult rats to the dietary saturated fat leads to ceramide accumulation in the tissues inducing ceramide synthesis from the palmitate and activating the SMases activities. It is rather the ceramide accumulation in the cells than neutral lipids or fatty acids accumulation is responsible and sufficient to inhibit the insulin-induced glucose uptake by the cells.

Keywords: Saturated fat; Palmitic acid; Ceramide; Sphingomyelin; Triacylglycerol; Diacylglycerol, Liver; Muscle; Brain cortex; Insulin

Introduction

The dietetic fatty acids are critical contributors to cell structure and function. However, excessive consumption of diets high in fats causes obesity and, subsequently, leading to chronic inflammation and disturbs the cellular metabolism. Dietetic fat-induced obesity increases the risk for insulin resistance, type 2 diabetes mellitus and neuro-degeneration. Non-esterified fatty acids have been implicated strongly in the development of skeletal muscle and liver insulin resistance and type 2 diabetes mellitus. Exposing muscle cells to palmitate inhibits insulin stimulation of a central mediator of insulin signaling in the target cells, serine/threonine kinase Akt/protein kinase B (PKB) (Chavez & Summers, 2003). This saturated fatty acid initiates the ceramide and diacylglycerol accumulation in insulin-resistant tissues. A high-fat lard-enriched diet causes a ceramide-associated hepatic insulin resistance and systemic inflammation, whereas a diet enriched by the unsaturated sunflower oil leads to the diacylglycerol-dependent insulin resistance development which is not reversed by the inhibition of sphingolipid synthesis (Holland et al., 2007). Among the persons with peripheral insulin resistance due to obesity, excess generation of toxic lipids, including ceramides, gives rise to brain insulin resistance, neuro-inflammation, and oxidative stress (de la Monte et al., 2009). The long-term high fat diet feeding was hypothesized to cause the excess hepatic production of neurotoxic ceramides that readily cross the blood-brain barrier and neuron death leading to Alzheimer's disease (AD)-like neuro-degeneration (Lyn-Cook et al., 2009). The precise physiologic mechanisms underlying these dietary influences are not completely understood but saturated fatty acid intake, as well as insulin resistance and neuro-inflammation likely contribute (Winocur & Greenwood, 2005; Parrott & Greenwood, 2007; Granholm et al., 2008; Nagle et al., 2009; de la Monte, 2009). A direct causal role of saturated fatty acids in the amyloidogenic processing of amyloid precursor protein and hyperphosphorylation of tau in neurons mediated by astroglia has been determined (Patil & Chan, 2005; Patil et al., 2006). At pathological levels of saturated fatty acids palmitic acid-induced de novo synthesis of ceramide in astroglia was found to play a critical role in increasing amyloid β production and hyperphosphorylating tau in neurons (Patil et al., 2007).

Sphingolipid ceramide is a novel class of bio-effectors molecules involved in the regulation of different signaling pathways. The factors that alter the metabolism of sphingolipids and ceramide production contribute significantly to progression of the pathogenesis of diabetes, atherosclerosis and neuro-degenerative disorders (for review see Cutler & Mattson, 2001). The ceramide accumulation in the hepatocytes is associated with inflammatory-like state in liver, which is manifested by increased production of the serum amyloid A, C-reactive protein and alpha-1-acid glycoprotein (Vranckx et al., 1995; Ballou et al., 1996) and interleikun 1beta (IL-1beta) hyperresponsiveness of the cells (Rurkute et al., 2007). Proinflammatory cytokine hyperresponsiveness was attributed to the age-dependent increase of neutral sphingomyelinase (SMase) acivity in the liver cells. Old animals fed with the calorie-restricted diet exhibited lower neutral sphingomyelinase acivity and reversed hyperresponsive phenotype (Rurkute et al., 2007), while the high fat diet increased the ceramide as well as triacylglycerol (TAG) production in the liver (Deevska et al., 2009). The palmitate flux through ceramide and TAG synthetic pathways is regulated in the liver in a competitive fashion and the capacity of liver to handle excess dietary fat and to accumulate TAG plays an important role in the regulation of overall glucose homeostasis. An important consequence of high fat diet-induced overproduction of ceramide in the liver was the increase of ceramide in brain structures and development of brain insulin resistance (Lyn-Cook et al., 2009). Ceramide accumulation in the neocortex, hippocampus and other subcortical regions of brain at AD neuropathology is associated with a diffuse loss of neurons and synapses (Puglielli et al., 2003; Cutler et al., 2004; Costantini et al., 2005). A clear relationship between sphingolipid metabolism altered and high saturated fat-enriched diet-induced cognitive impairment has been determined (Babenko et al., 2009).

In the chapter the short- and long-term effects of dietary saturated fats and palmitic acid on ceramide and neutral lipids accumulation in the insulin responsive tissues and insulin-induced glucose uptake have been considered. Two-week long treatment of male rats with dietetic beef fat increased sphingomyelin (SM) and decreased the total phosphalipids (PL) contents in the liver. Short-term saturated fat-enriched diet had no effect on the free fatty acids contents and neutral lipids and glycerolphosholipids production in the liver and other tissues studied, while the long-term feeding of rats with high fat diet significantly increased the ceramide synthesis in the liver, brain cortex and diaphragm muscle. Tissue specific effects of dietetic saturated fat on sphingomyelin and neutral lipids synthesis have been determined. Exposure of hepatocytes, diaphragm strips and cortex slices of adult rats, having a free access to a standard chow diet, to palmitate or short-chain ceramide increased significantly the content of the newly synthesized ceramide and decreased the insulin-induced glucose uptake by the cells. These findings suggest the ceramide is a sufficient to inhibit insulin signaling in the classical target tissues and brain.

Materials and Methods

Materials

[^{14}C]palmitic acid (56 mCi/mmol) – Amersham, GE Health Care UK (UK), [^{14}C]serine (25 mCi/mmol) – BPO Isotop (Russia); silica gel plates for thin-layer chromatography from Sorbfil (Russia). D-erythro-sphingosine, N-acetyl (C2-ceramide), purity 98 %, was from Alexis (Grünberg, Germany). Lipid standards (ceramide, SM, phosphatidylcholine (PC), triacylglycerol (TAG), diacylglycerol (DAG), palmitic acid) were obtained from Sigma-Aldrich. Other chemicals used were of chemically pure grade.

Animals

In the experiments the 3-month-old male Wistar rats were used. They were kept at 24°C on a cycle of 12 h light/12 h darkness. Experimental procedures were approved by the Institutional Animal Care and Use Committies at the Kharkov Karazin National University. The animals were divided into two groups. They were fed by a standard chow diet (control group) (14% fat-derived calories, 12% protein, 75% carbohydrate) or beef fat (experimental group) (32% fat-derived calories, 9% protein, 59% carbohydrate). Rats of the experimental and control groups were fed by the diets for 2, 4 or 36 weeks. They were fasted overnight and were anesthetized the next day with an injection of ketamine (75 mg/kg) and sacrificed by decapitation. Liver, diaphragm and brain cortex were isolated and used for preparation of homogenates and tissue slices and lipid analysis as described below. 3-month-old rats, which had a free access to a standard chow diet, were used for hepatocyte isolation as described below.

Experiments on Liver, Diaphragm and Brain Cortex

The liver was perfused with 0.9% NaCl, then removed and washed in Krebs-Henseleit buffer, pH 7.4, containing 2 mM $CaCl_2$ and 0.2% BSA. Brains were quickly removed, dissected on ice and cortex was isolated and washed in Krebs-Ringer bicarbonate buffer, pH 7.4. The entire diaphragm was dissected, trimmed of fat and washed in Krebs-Henseleit buffer, pH 7.4, containing 2 mM $CaCl_2$ and 0.2% BSA. Tissues slices were prepared in Krebs-Ringer bicarbonate buffer, pH 7.4, and used to determine the [^{14}C]palmitic acid intake and sphingolipids and neutral lipids synthesis as described below. The lipids were extracted and analyzed as described below. To study the effect of exogenous palmitic acid or ceramide on tissues, brain cortex slices and diaphragm strips were incubated in Krebs-Ringer bicarbonate buffer, pH 7.4, in the presence of palmitic acid (0.75 mM), bound to fatty acid-free bovine serum albumin (1:1, molar ratio) or C2-ceramide (10μg/ml) for 6 h in a humidified atmosphere of 5% CO_2 in a tissue culture incubator at 37°C followed by incubation in the absence or presence of insulin (100 nM) during the last 10 min of incubation with palmitic acid or C2-ceramide. Control tissues were incubated in the presence of fatty acid-free bovine serum albumin/ethanol. Glucose uptake was assayed by incubation with 1 μCi/ml 2-deoxy-D-[3H]glucose for 10 min as described previously (Hajduch et al., 1998; Blair et al., 1999).

Experiments with Hepatocytes

Hepatocytes were isolated from the 3-month-old male Wistar untreated rats by the method described in (Gustavsson et al., 1994). Preparation of hepatocytes was started between 9:00 and 10:00 a.m. The cells (10^7/ml) were incubated in Eagle medium containing 10% fetal calf serum, 100 units/liter streptomycin, 100 units/liter penicillin in the presence of palmitic acid (0.75 mM), bound to fatty acid-free bovine serum albumin (1:1, molar ratio) or C2-ceramide (10μg/ml) for 6 h in a humidified atmosphere of 5% CO_2 in a tissue culture incubator at 37°C. Control cells were incubated in the presence of fatty acid-free bovine serum albumin/ethanol. The Petri dishes coated with a solution of rat tail collagen were used (Michalopoulos & Pitot, 1975). The cells were released from the dishes by a mild trypsin treatment. The hepatocytes were allowed to settle and the pellet was resuspended in the Krebs-Henseleit buffer, pH 7.4, containing 2 mM $CaCl_2$, 25 mM HEPES, 0.1% BSA. To determine the cells viability, the trypan blue exclusion into hepatocytes was studied. Before lipid extraction, the cells were washed twice with the Krebs-Henseleit buffer, pH 7.4, containing 2 mM $CaCl_2$, 25 mM HEPES, 0.1% BSA. The lipids (SM, ceramide, neutral lipids) were extracted and analyzed as described below.

Determination pf Sphingolipids and Neutral Lipids Synthesis

The tissues homogenates were incubated in Krebs-Ringer bicarbonate buffer, pH 7.4, containing 20μCi/ml [^{14}C]palmitic acid, for 90 min. Reaction was terminated by the addition of chloroform/methanol (1:2, v/v). Lipids were extracted and newly synthesized TAG, DAG, SM and ceramide analyzed as described below. The isolated hepatocytes were incubated in the Eagle medium containing 10% fetal calf serum, 100 units/liter streptomycin, 100 units/liter penicillin, 13 mg/ml gentamycin, palmitic acid (0.75 mM) bound to fatty acid-free bovine serum albumin (1:1, molar ratio) or C2-ceramide (10μg/ml) and 20 μCi/ml [^{14}C]palmitic acid or [^{14}C]serine for 6h. Control cells were incubated in the presence of fatty acid-free bovine serum albumin/ethanol. Reaction was terminated by the addition of chloroform/methanol (1:2, v/v). Lipids were extracted and newly synthesized TAG, DAG, SM and ceramide analyzed as described below.

Extraction and Separation of Lipids

The lipids were extracted according to the Bligh & Dyer (1959) protocol. The chloroform phase was collected and dried under N_2 at 37°C. The lipids were re-dissolved in chloroform/methanol (1:2, v/v) and applied on TLC plates. For TAG, DAG and free fatty acids separation – hexane/diethyl ether/acetic acid (80:20:2, v/v) was used. For PC separation, the solvent system: chloroform/methanol/acetic acid/water (25:15:4:2, v/v) was used. For sphingolipids analyses dried lipids were incubated in the chloroform/methanol (1:1, v/v) containing 0.1M KOH for 60 min for acylglycerolipid hydrolysis. Lipids were re-extracted (Bligh & Dyer, 1959) and dried. For SM and ceramide separation, the plates with loaded lipid extract were developed with chloroform/ethylacetate/izopropylanol/methanol/0.25 % KCl (25:25:25:10:9, v/v). The appropriate standards were applied on each plate for quantification. The regions migrating with the standart ceramide were scraped from the plate and the ceramide eluted from the silica with 1 ml chloroform: methanol (1:1, v/v) followed by 1 ml of methanol. The combined eluates were dried in vacuo, and the ceramide mass was quantified by long chain bases released after an acid hydrolysis in 0.5 M HCl in methanol at 65°C for 15 h (Sathishkumar et al., 2005). Free long chain bases were analyzed as described by Lauter & Trams (1962). The contents of total PL, SM, and PC in chromatographic fractions were determined by the method of Bartlet (1966) and TAG and DAG – by the method of Marsh & Weinstein (1966). The gel spots containing [^{14}C]lipids were scraped and transferred to scintillation vials. Radioactivity was measured by a scintillation counter. Hepatic fatty acid composition was analyzed by gas chromatography after trans-methylation using C17:0 as internal standard (Muskiet et al., 1983).

Statistical Analysis

One-way analysis of variance (ANOVA) procedures was used to assess significant differences among treatment groups. Student's t-test was used for paired observations and significance was set at $p < 0.05$ and $p < 0.001$.

Results and Discussion

High-fat diets are well-known to increase body weight, body fat and induce insulin resistance in both animals and humans. Dietetic saturated fats can rapidly increase liver fat levels as well as fasting plasma insulin and glucose levels and are associated with the hepatic insulin resistance (Samuel et al., 2004). The accumulation of hepatic TAG depends on the fat saturation and its content in the diet and animal model used in the experiments. A short-term feeding of rats by saturated fat-enriched diet had no effect on body weight, whereas, after 4- or 36 weeks of fat feeding, body weight significantly increased as compared with control rats (Fig. 1). Although rats fed a saturated fat-enriched diet for 2 weeks had body weights similar to those of control animals, the SM content was higher and total phospholipids and PC contents were lower in their livers (Fig. 2A). 2-weeks feeding of rats with the saturated fat did not change the levels of the free fatty acids and TAG (Fig. 2A) and total PL and TAG synthesis in the liver (Fig. 2B) and other tissues studied (data not shown). However, 2-weeks feeding of rats with the saturated fat increased the level of palmitic (16:0) and oleic acids (18:1) and reduced the arachidonic acid (20:4) in the liver (Table 1). The results obtained are in line with the experiments on short-term (3 – 28-day-long) feeding of mice with the saturated palm oil (45 kcal fat) (de Wilde et al., 2008). Moreover, upon a short-term dietary fat supplementation of human subjects de novo lipogenesis was not induced (Schwarz et al., 1995; Hudgins et al., 1996). It demonstrated that in the skeletal muscle the short-term saturated high-fat diet did not affect the total body weight and the contents of free fatty acids, TAG, DAG and total glycerolphospholipids. However, saturated dietetic fat increased the content of saturated fatty acids (mainly the palmitic one) in the membranes phospholipids (de Wilde et al., 2008). Palmitate, which content increased in the liver under dietetic saturated fat action, could be elongated to stearate (18:0) and converted to oleate (18:1) (Oosterveer et al., 2009). A high-fat diet containing beef fat significantly increased the expression of stearoyl-CoA desaturase and oleic acid (18:1) content in the mice liver. Taking into account that total contents of saturated and unsaturated fatty acids in the liver lipids did not differ between saturated fat fed and control rats (Table 1), it could not be excluded that dietetic fat-induced increase of the oleate synthesis is the adaptive reaction of the liver cells that allows to maintain stable saturation of lipids which are of great importance for membrane fluidity and thus for normal membrane function.

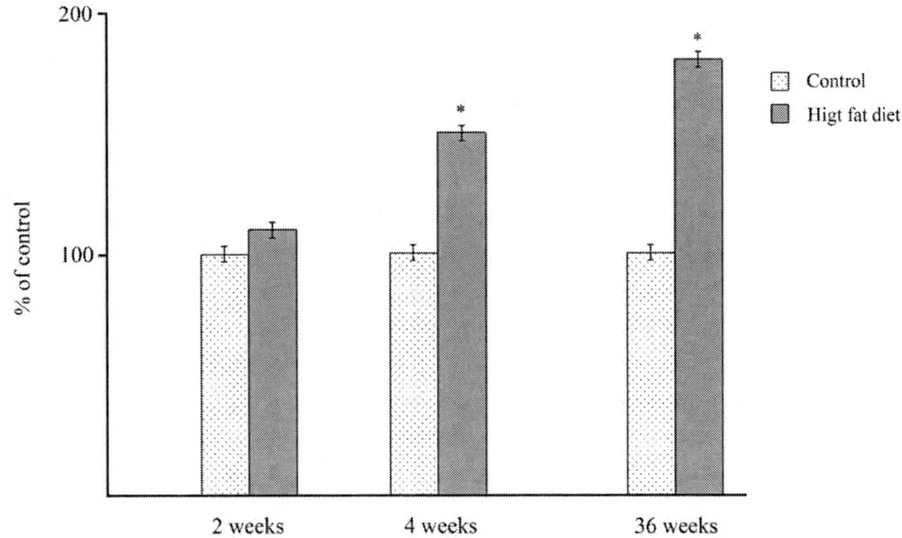

Figure 1. Body weight of control and saturated fat treated rats. The animals were fed by a standard chow diet (control group) (14% fat-derived calories, 12% protein, 75% carbohydrate) or beef fat (experimental group) (32% fat-derived calories, 9% protein, 59% carbohydrate) for 2, 4 or 36 weeks. Control animals were fed by a standard chow diet and were the same age as 2- or 4- or 36-week-long saturated fat enriched-diet-treated rats. There were ten animals in each group. * $P<0.05$, saturated fat-treated vs. control.

Palmitic acid is the precursor of sphingolipid synthesis de novo and dietetic saturated fat could increase the serine palmytoil transferase expression in the liver and adipose tissues (Lyn-Cook et al., 2009; Shah et al., 2008). The excess of palmitic acid content in the diet is supposed to intensify the sphingolipid synthesis in the liver. However, 2-week-long saturated fat-enriched diet increased the SM and decreased the PC contents and did not change significantly the level of the ceramide in the liver (Fig. 2). Thus, in response to increased saturated fatty acid supply, the liver exerts several adaptive physiological reactions to prevent accumulation in the cells of cytotoxic lipid species. It is possible that under short-term action of dietetic saturated fat the excesses of the free saturated fatty acid palmitate could be targeted mainly into SM and oleic acid synthesis due to SM synthase and stearoyl-CoA desaturase activation, respectively.

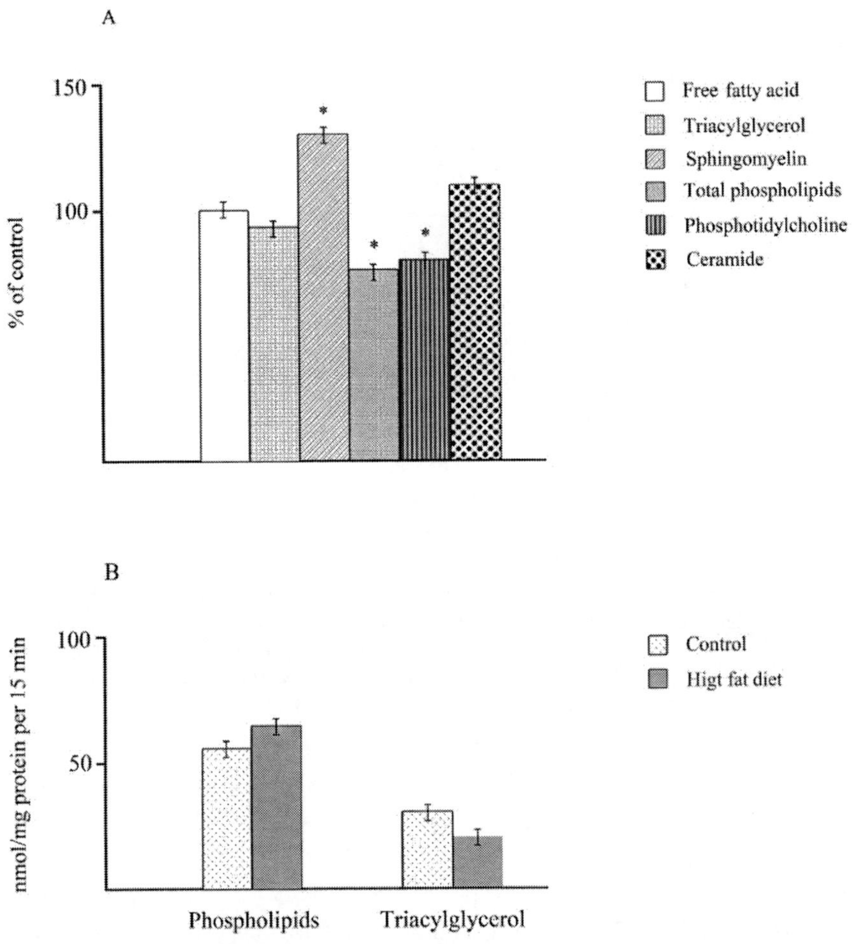

Figure 2. Effect of short-term saturated fat-enriched diet on the lipids contents and synthesis in the liver. The animals were fed by a standard chow diet (control group) (14% fat-derived calories, 12% protein, 75% carbohydrate) or beef fat (experimental groups) (32% fat-derived calories, 9% protein, 59% carbohydrate) for 2 weeks. A. Liver lipids contents. B. Liver lipids synthesis. Liver homogenates were used to study the lipid synthesis as described in the Section: Materials and Methods. Results are mean ± SE of ten experiments performed in duplicate. * $P<0.05$, saturated fat-treated vs. control.

Table 1. Effects of short-term saturated fat-enriched diet on fatty acid composition of liver total lipid

Fatty acid	Control	Saturated fat
12:00	0.32 ± 0.03	0.52 ± 0.04
14:00	0.20 ± 0.02	0.26 ± 0.03
15:00	0.16 ± 0.02	0.22 ± 0.03
16:00	24.2 ± 0.84	29.9 ± 0.40*
16:1n-9	1.58 ± 0.12	1.10 ± 0.07
17:00	0.91 ± 0.11	0.87 ± 0.04
17:01	0.54 ± 0.11	0.49 ± 0.09
18:00	28.3 ± 0.66	26.9 ± 0.30
18:1n-9	17.4 ± 0.49	21.3 ± 0.20*
18:2n-6	9.90 ± 0.27	10.2 ± 0.23
18:3n-3	0.73 ± 0.13	0.77 ± 0.12
20:4n-6	15.2 ± 0.51	10.8 ± 0.17*
Saturated	53.7 ± 2.59	55.7 ± 0.76
Monounsaturated	19.5 ± 1.74	22.9 ± 0.53
Polyunsaturated	25.8 ± 1.86	21.8 ± 0.93

The animals were fed by a standard chow diet (control group) (14% fat-derived calories, 12% protein, 75% carbohydrate) or beef fat (experimental group) (32% fat-erived calories, 9% protein, 59% carbohydrate) for 2 weeks. Tissues homogenates were used to study fatty acid composition as described in the Section: Materials and Methods. Results are mean ± SE of ten experiments performed in uplicate. * $P<0.05$, saturated fat-treated vs. control.

In contrast to the short-term saturated fat-enriched diet the long-term fatty diet led to the significant changes in the neutral lipids and ceramide synthesis in the tissues. 4-weeks and 36-weeks feeding of rats with the saturated fat increased the [^{14}C]fatty acids levels, TAG and DAG synthesis in the liver, diaphragm and brain cortex tissues (Fig. 3). The more pronounced effect of the experimental diet has been found in the diaphragm (Fig. 3B) as compared with the liver and brain (Fig. 3A, C). The effects of saturated fat on fatty acids and neutral lipids synthesis in the diaphragm muscle were highly dependent on the diet longevity used. 4-week-long diet treatment led to more than twice increase of [^{14}C]fatty acids level in the diaphragm (Fig. 3B), although the newly synthesized TAG content increased by 210% with respect to control. 36-week-long saturated fat-enriched diet increased the TAG synthesis in the diaphragm by 70% and had no effect on the newly synthesized fatty acids as compared with control diet-treated rats.

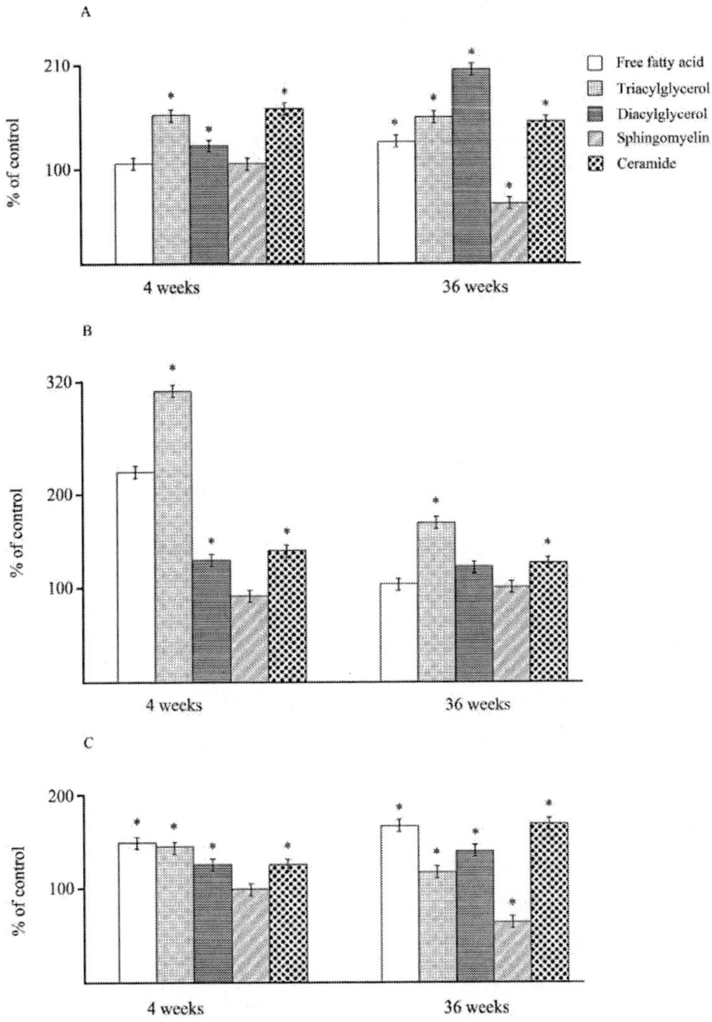

Figure 3. Effect of long-term saturated fat-enriched diet on the lipids synthesis in the liver, diaphragm and brain cortex tissues. The animals were fed by a standard chow diet (control group) (14% fat-derived calories, 12% protein, 75% carbohydrate) or beef fat (experimental groups) (32% fat-derived calories, 9% protein, 59% carbohydrate) for 4 or 36 weeks. A. Liver. B. Diaphragm muscle. C. Brain cortex. Tissues homogenates were used to study the lipid synthesis as described in the Section: Materials and Methods. Results are mean ± SE of ten experiments performed in duplicate. * $P<0.05$, saturated fat-treated vs. control.

The long-term high-fat diet induces the lipogenic gene expression (Lin et al., 2005; Sampath et al., 2007). But it does not result in a significant induction of de novo lipogenesis in mice and rats (Delgado et al., 2009; Oosterveer et al., 2009).

Expression of genes encoding enzymes involved in hepatic TAG synthesis (glycerol-3phosphate acyltransferase and diacylglycerol acyltransferase) has been observed in animals fed the high-fat diet compared to animals receiving the low-fat diet (Osterveer et al., 2009). Moreover, the long-term saturated fat-enriched diet increases the elongation of palmitic acid and fatty acid desaturation and does not affect the hepatic VLDL-TAG secretion. The above results indicate that long-term saturated fat-enriched diet stimulates fatty acid synthesis from chain elongation and desaturation and neutral lipids synthesis and thus increases the level of newly synthesized fatty acids, DAG and TAG in the liver and other tissues studied.

Elevated intracellular fatty acids levels are thought to play a causal role in hepatic and skeletal muscle insulin resistance. However, the redirection of fatty acids into TAG storage may be the initial protective mechanism to lower intracellular fatty acid concentration. Although insulin resistance correlates highly with TAG content in skeletal muscle and liver, TAG itself does not disrupt the insulin signaling in the target cells. It is noteworthy, that the excess of the de novo synthesized DAG activates the protein kinase (PKC), phosphorylation of insulin receptor substrate-1 (IRS-1), which suppresses the IRS-1 tyrosine phosphorylation and diminishes phosphatidylinositol 3-kinase activation and thus blocks the insulin signaling in the target cells (Morino et al., 2006; Schmitz-Peiffer & Biden, 2008; Neschen et al., 2005; Nagle at al., 2007; Benoit et al., 2009). The results and data presented in the Chapter suggest that the chronically elevated de novo DAG synthesis in the muscle, brain and liver tissues induced by the dietetic saturated fat underlying in defective insulin-stimulated glucose transport in the cells. As both the saturated and unsaturated fatty acids cause insulin resistance and the PKC activation is relatively specific to "polyunsaturated" DAG species (Schmitz-Peiffer & Biden, 2008) one can conclude that the dietetic saturated fat does not act specifically via the DAG/PKC pathway. It is not always possible to explain the diminished insulin signaling by the DAG oversupply and PKC activation. Other lipid intermediates have been recently implicated as mediators of the saturated fat-induced insulin resistance.

The saturated fatty acids oversupply generates ceramide and sphingolipid accumulation in the cells. The long-term studies revealed that dietetic saturated fat significantly increased the level of the newly synthesized ceramide in the liver, diaphragm muscle and brain cortex tissues (Fig. 3). The dietetic saturated fat increased the content of ceramide in the tissues studied. Ceramide contents in the liver of 4-week-long diet-treated and control rats were 8.80 ± 0.45 and 6.00 ± 0.19 nmol/mg protein ($P<0.05$, saturated fat-treated vs. control), respectively. Ceramide contents in the brain cortex of diet-treated and control animals were

19.4 ± 1.02 and 11.7 ± 0.80 nmol/mg protein ($P<0.05$, saturated fat-treated vs. control), respectively. Thus, the long-term saturated fat treatment of rats due to the sphingolipid synthesis stimulation could increase the content of ceramide in insulin responsive tissues.

Several studies have demonstrated that increased supply of palmitic acid enchances de novo synthesis of sphingolipids, including ceramide, in the liver cells, myotubes (Merrill et al., 1995; Chavez & Summers, 2003) and cortex astroglia (Patil et al., 2007). The long-term saturated fat-enriched diet where 42 - 60% of total calories were derived from fat significantly increased the ceramide content in mice blood plasma, adipose tissue, liver and liver cell nuclei (Shah et al., 2008, Lyn-Cook et al., 2009, Chocian et al., 2010; Boini et al., 2010). To determine the mechanism of the dietetic saturated fat-induced ceramide accumulation in the cells, the gene expression of enzymes involved in ceramide production and turnover has been investigated. High fat-diet significantly increased the expression and activity of the enzymes involved in the ceramide synthesis (serine palmitoiltransferase and ceramide synthase) in mice liver and adipose tissues, but not in the brain (Shah et al., 2008, Lyn-Cook et al., 2009). As compared with the low-fat-diet the long-term high fat-diet enhanced significantly ceramide synthesis de novo in the liver (Fig. 1A), diaphragm muscle (Fig. 3B) and brain cortex (Fig. 3C) tissues of adult rats. Moreover, the 36-week-long treatment of the rats with the dietetic saturated fat decreased the level of the newly synthesized SM in the liver and brain cortex as compared with the control animals, while the level of [^{14}C]SM and [^{14}C]ceramide/SM ratio (Table 2) in the liver, diaphragm muscle and brain cortex tissues of rats treated by saturated fat for 4 weeks did not significantly differ from that of the animals received the low fat diet. 4-week-long saturated fat treatment of mice had no effect on liver acid and neutral SMase activities and colon neutral SMase activity as compared with low fat diet-treated animals (Cheng et al., 2004). 36-week-long treatment of adult rats with dietetic saturated fat significantly increased the [^{14}C]ceramide/SM ratio in the liver and brain cortex with respect to control ones (Table 2). Taking into account that the altered ceramide/SM ratio may itself be a consequence of the elevated SMase activity, our findings suggest that saturated fat chronically added to the diet of rats could increase the SMase activity in the liver and brain.

Table 2. Effects of saturated fat-enriched diets on [^{14}C]ceramide/sphingomyelin ratio in the liver, diaphragm muscle and brain cortex tissues of rats

Animals	Liver	Diaphragm	Cortex
Control[a]	1.63 ± 0.08	0.56 ± 0.09	1.18 ± 0.04
4-week-long fat enriched-diet	1.52 ± 0.06	0.95 ± 0.16	1.56 ± 0.08
Control[b]	1.03 ± 0.04	1.33 ± 0.07	0.65 ± 0.05
36-week-long fat enriched-diet	3.28 ± 0.07*	1.75 ± 0.14	2.33 ±0.07*

The animals were fed by a standard chow diet (control group) (14% fat-derived calories, 12% protein, 75% carbohydrate) or beef fat (experimental group) (32% fat derived calories, 9% protein, 59% carbohydrate) for 4 or 36 weeks. Control[a] and Control[b] rats are the animals that were fed by a standard chow diet and were the same age as 4- and 36-week-long saturated fat enriched-diet-treated rats, respectively. Tissues homogenates were used to study the lipid synthesis as described in the Section: Materials and Methods. Results are mean ± SE of ten experiments performed in duplicate. * $P<0.05$, saturated fat-treated vs. control.

Long-term saturated fat feeding of the mice led to the pronounced increase of SMases expression and activity in the liver and the epididymal and subcutaneous fat pads (Shah et al., 2008, Lyn-Cook et al., 2009). However, high fat diet induced the expression of both neutral and acid SMases in the fat tissues (Shah et al., 2008), increased activity of neutral SMase but not acid SMase in the liver and liver cell nuclei (Chocian et al., 2010) and activated acid SMase but not neutral SMase in the colon (Cheng et al., 2004). These findings and results presented in the Chapter suggest that dietetic saturated fat-induced expression of SMases mRNA can be the important reason of chronic sustained ceramide accumulation in the different tissues. Saturated fat-induced activation of different types of SMases is highly tissue specific and depends on the length of dietetic action on animal.

Dietetic saturated fat-induced ceramide accumulation in the blood plasma and insulin responsive tissues is associated with the altered glucose homeostasis and development of the insulin resistance. Long-term saturated fat enriched diet led to the sustained increased glucose and insulin levels in the fasted state in the blood plasma (Deevska et al., 2009; Boini et al., 2010; Shah et al., 2008). Insufficient insulin-mediated suppression of plasma glucose after insulin administration to saturated fat-treated mice has been found. The decreased high fat-induced expression of serine palmitoiltransferase, neutral and acid SMase and ceramide content in the blood plasma and adipose tissue in plasminogen activator inhibitor-1$^{-/-}$ mice ameliorate insulin resistance (Chah et al., 2008). Treatment of mice with an acid SMase inhibitor, amitriptyline, significantly attenuated the diet-induced plasma ceramide level, improved glucose homeostasis and prevented development of the insulin resistance (Boini et al., 2010). However, inhibitor of ceramide synthesis de novo, myriocin, administered to the saturated fat-treated Sprague-Dawley rats or Zucker diabetic fatty rats decreased the ceramide contents in the blood serum, soleus muscle and liver and improved glucose homeostasis (Holland et al., 2007).

A majority of studies evaluating the mechanisms of sphingolipid-induced insulin resistance have employed classical insulin target tissues including skeletal muscle, fat and liver. Ceramide addition to the cultured myotubes, adipocytes or hepatocytes inhibited glucose uptake or glycogen synthesis via inhibition of Akt/PKB (Holland & Summers, 2008). Ceramide blocked the translocation of Akt/PKB to the plasma membranes (Stratford et al., 2001) and stabilized interactions between Akt/PKB and atypical PKC isoform PKCζ (Hajduch et al., 2008; Fox et al., 2007). Phosphorylation of serine 34 of the Akt/PKB PH domain blocked its ability to interact with 3-phosphoinositide kinase (PI3K). Ceramide activating the protein phosphatase 2A (PP2A) could dephosphorylate Akt/PKB and prevent insulin stimulation of glucose uptake (Ugi et al., 2004). The PP2A inhibitor okadaic acid added to the cultures of C2C12 myotubes (Chavez et al., 2003), PC12 neurons (Salinas et al., 2004), adipocytes (Teruel et al., 2000), or human glioblastoma cell line (Zinda et al., 2001) prevented the disruption of insulin signaling pathway by the ceramide. In addition to ceramide early effects on insulin signaling other mechanisms were supposed. Low doses of ceramide inhibited glucose uptake via Akt/PKB-independent pathway, attenuating GLUT4 translocation (JeBailey et al., 2007). Moreover, ceramide could decrease GLUT4 transcription and induce the insulin resistance in 3T3-L1 adipocytes (Long & Pekala, 1996).

Addition of the C2-ceramide to the culture media of isolated hepatocytes as well as diaphragm strips and cortex slices of adult rats led to the accumulation of endogenous long-chain ceramides and had no effect on the TAG, DAG and free fatty acids contents in the cells (Fig. 4). Due to their relatively high solubility in aqueous media, exogenous, short-chain ceramides rapidly penetrate to the cells and convert into long-chain ceramides (Ogretmen et al., 2002; Le Stunff et al., 2007). 3-hour-long incubation of FL5.12 cells in the presence of C2-ceramide increased the cellular contents of total long-chain ceramides, C16-, C18- and C20-ceramides, and later also downregulated the glucose transporter GLUT-1

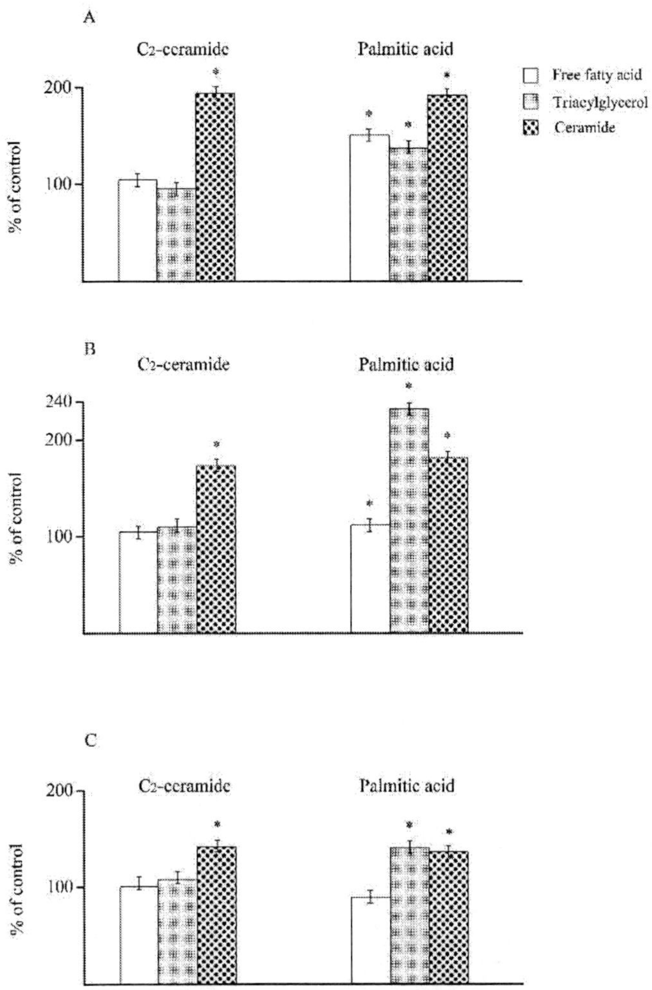

Figure 4. Effect of C2-ceramide and palmitic acid on lipids synthesis in the liver cells, diaphragm and brain cortex tissues. A. Liver. B. Diaphragm muscle. C. Brain cortex. To study C2-ceramide effects on the contents of the newly synthesized lipids, the isolated hepatocytes, brain cortex slices and diaphragm strips were incubated in the presence of C2-ceramide (10μg/ml) for 6 h. Control cells and tissues were incubated in the presence of ethanol. To study palmitic acid effects on the newly synthesized lipids contents, the isolated hepatocytes, brain cortex slices and diaphragm strips were incubated in the presence of palmitic acid (0.75 mM), bound to fatty acid-free bovine serum albumin (1:1, molar ratio) for 6 h. Control cells and tissues were incubated in the presence of fatty acid-free bovine serum albumin. Hepatocytes isolation, tissue slices preparation and culturing and lipid synthesis and analysis were carried out as described in the Section: Materials and Methods. Results are mean ± SE of ten experiments performed in duplicate. * $P<0.05$, palmitic acid- or C2-ceramide-treated vs. control.

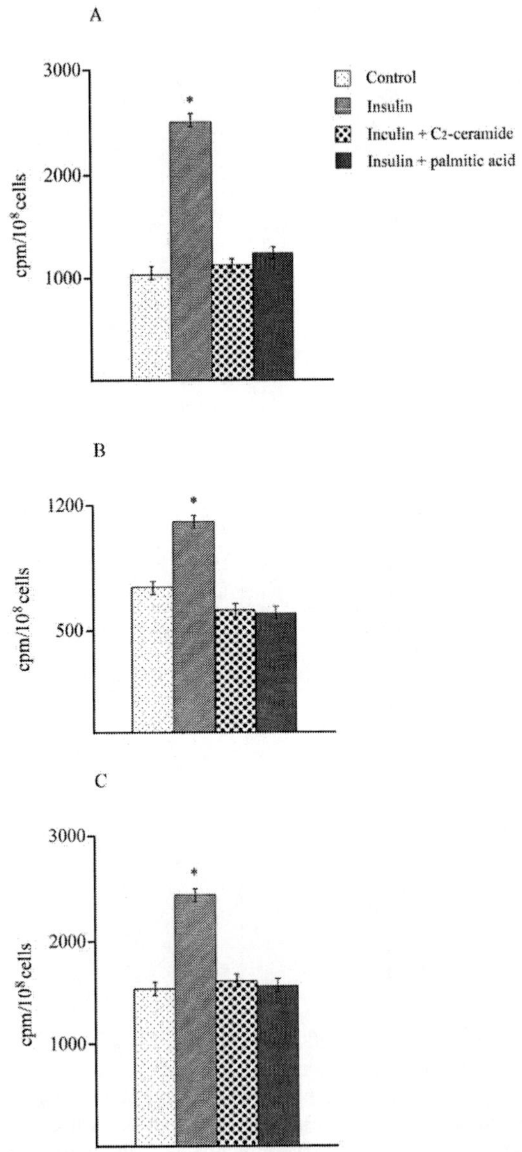

Figure 5. Effect of C2-ceramide and palmitic acid on insulin stimulated glucose uptake by the liver cells, diaphragm and brain cortex tissues. A. Liver. B. Diaphragm muscle. C. Brain cortex. To study C2-ceramide and palmitic acid effects on the insulin stimulated glucose uptake by the cells used rats, which were fed by a standard chow diet (14% fat-derived calories, 12% protein, 75% carbohydrate). Hepatocytes isolation, tissue slices preparation and culturing and glucose uptake were made as described in the Section: Materials and Methods. Results are mean ± SE of ten experiments performed in duplicate. * $P<0.05$, control vs. insulin.

(Guenther et al., 2008). Studies on different cell cultures have demonstrated the efficient hydrolysis of C2-ceramide to sphingosine. Sphingosine generated by micelle-derived C2-ceramide efficiently re-acylated to long-chain ceramides and increased cellular ceramides content (Boyanovsky et al., 2003). Our observations that the cell-permeant C2-ceramide increased contents of cellular long-chain ceramides and downregulated insulin-induced glucose uptake by the cells (Fig. 5), strongly suggest that not only the extracellularly added ceramide, but the generated in the cells ceramide may be involved in the inhibition of insulin signaling in the target cells.

It was well documented that palmitic acid, added to the culture media or infused to the animals, could mimic some effects of short- and long-chain ceramides in the insulin signaling events. Infusion of the lard oil, enriched by the palmitic acid, to the rats inhibited whole body 2-deoxyglucose uptake, decreased the glucose infusion rate and increased hepatic glucose output (Holland et al., 2007). Moreover, it increased the ceramide and DAG contents in soleus muscles and inhibited insulin stimulation of Akt/PKB. Inhibition of ceramide synthesis by myriocin led to decreased ceramide content in soleus muscles and prevented insulin resistance development in lard-infused animals versus vehicle controls. The similar results were obtained on the palmitate-treated muscle strips (Holland et al., 2007) and cultured myotubes (Chavez & Summers, 2003; Schmitz-Peiffer et al., 1999).

Culturing of the liver cells, diaphragm strips and cortex slices in the presence of palmitic acid increased the contents of neutral lipids (TAG, DAG), free fatty acids and ceramide in the cells and tissues studied (Fig. 4). However, the palmitic acid and also C2-ceramide reduced the [^{3}H]-D-glucose intake by the cells stimulated by insulin (Fig. 5). It can be assumed from these findings that ceramide accumulation in the cells rather than accumulation of the neutral lipids or fatty acids is sufficient for the inhibition of insulin-induced glucose uptake by the cells and brain tissue is extremely sensitive to exogenous palmitate and ceramide. Studies in vivo supported the idea that ceramide accumulated in the insulin responsive rat tissues, mediates, at least in part, saturated fat-induced insulin resistance development. Saturated fatty acid-induced synthesis of the neutral lipids could prevent the oversupply of the free fatty acids and ceramide and thus protect the cells against death.

Human epidemiologic studies provided convincing evidence that hyperinsulinemia as well as type II diabetes mellitus are among the risk factors for AD (Parrott & Greenwood, 2007; Neumann et al., 2008). Peripheral hyperinsulinemia correlates with an abnormal removal of the amyloid beta peptide and an increase of tau hyperphosphorylation that leads to decline in cognitive function. Diets high in saturated fats as well as type II diabetes mellitus modulate brain insulin activity and lead to the cognitive decline. Previously we reported that the enrichment of the diet of old Wistar rats with saturated beef fat increased significantly ceramide and sphingosine levels in the brain cortex and ceramide content in the hippocampus (Babenko & Semenova, 2010). Diet- and age-induced disturbances of sphingolipid turnover in the brain were associated with microglia activation, neuroinflammation, loss of hippocampal dendritic integrity and cognitive decline (Granholm et al., 2008; Babenko et al., 2009; Babenko & Semenova, 2010). Elevation of ceramide content was observed in the cortex and hippocampus (AD-affected regions) of aged animals and at AD pathology (Cutler et al., 2004). AD-like pathophysiological changes in the brain could be induced by palmitic acid and increased ceramide levels have been suggested to be central to AD pathology. Exogenous short-chain ceramide along with palmitic acid or neutral SMase treatment increases the beta-secretase BACE-1 half-life and thereby promotes amyloid beta biogenesis in the cortical neurons which have been found to be mediated by astroglial lipid metabolism (Puglielli et al., 2003, Patil et al., 2007, 2008). Pre-treatment of the neurons with amyloid beta oligomers stimulated acid SMase and ceramidase activites, and ceramide and sphingosine accumulation in the cells (He et al., 2010). The data presented in the Chapter have demonstrated that dietetic saturated fat can induce the newly synthesized ceramide accumulation in the brain cortex of adult rats. The saturated fat-induced activation of sphingolipid metabolism in the brain could be imitated by the exogenous palmitic acid or ceramide addition to the culture media of brain cortex slices. Our results suggest that exposure of adult rats to the dietary saturated fat induced ceramide accumulation in the brain cortex via two different mechanisms: 1) ceramide synthesis from the palmitate, and 2) increased ceramide content in the liver and sphigolipid transport to the brain. Ceramide accumulated in cortex can play the central role in saturated fat-induce brain insulin resistance.

Conclusion

We demonstrated that short-term as well as long-term saturated fat enriched diet significantly affects the lipids metabolism. Attention has been focused on ceramide, SM and neutral lipids synthesis in insulin responsive tissues. We have shown that effect of dietetic saturated fat on lipid turnover is tissue-specific and time-dependent. Short exposure of rats to saturated dietetic fat increases SM and decreases PC levels and does not change the ceramide and neutral lipids and free fatty acids contents in the liver. However, short-term diet increased palmitic and oleic acids and decreased arachidonic acid contents in the liver. Therefore, one can conclude that with the short-term excess of saturated fat in the diet palmitic acid mainly challenged to the SM synthesis and desaturation/elongation processes. The saturated fat-induced activation of unsaturated fatty acids turnover in membrane phospholipids could maintain the stable saturation of the membranes. The long-term treatment of rats by dietetic saturated fat led to accumulation of the newly synthesized ceramide, neutral lipids and fatty acids in all tissues studied and had no effect on the SM synthesis. Chronic (36-week-long) feeding of rats by the saturated fat decreased significantly the newly synthesized SM levels and increased the ceramide/SM ratio in the liver and brain. The suggestion has been made that very long chronic treatment of rats by the dietetic saturated fat could enhance not only the ceramide synthesis de novo but its production via SMase activation and the newly synthesized SM degradation in the liver and brain cortex. It is noteworthy that the long-term saturated fat-enriched diet led to the more pronounced effect on the newly synthesized ceramide accumulation in brain cortex and liver than in muscle. However, in the diaphragm muscle dietetic palmitic acid mainly challenged to the TAG synthesis. Effects of dietetic saturated fat on ceramide and neutral lipids synthesis in the insulin target tissues could be mimicked by palmitic acid in experiments in vitro. Exogenous short-chain ceramide simulated the saturated fatty acid effect on long-chain ceramide synthesis and insulin-stimulated glucose uptake by the cells and did not act on fatty acids and neutral lipids accumulation. These results strongly suggest that at the patholocical levels of palmitic acid fatty acid could be metabolized into sphingo- and glycerolipids while in the classical insulin target cells and in the brain cortex, too the cellular ceramide accumulation was required for the insulin resistance development. Under pathological chronically elevated level of saturated fatty acids brain may be affected by abnormal fatty acid metabolism leading to the overproduction of endogenous brain ceramide and brain insulin resistance.

References

Babenko, NA; Semenova, YaA; Kharchenko, VS. Effects of fat-enriched diet on the content of sphingolipids in the brain and on cognitive functions in old rats. *Neurophysiology*, 2009 41, 258-263.

Babenko, N. A. & Semenova, Ya. A. (2010). Effects of long-term fish oil-enriched diet on the sphingolipid metabolism in brain of old rats. *Exp Gerontol*, 45, 375-380.

Ballou, SP; Lozanski, FB; Hodder, S; Rzewnicki, DL; Mion, LC; Sipe, J; Ford, AB; Kushner, I. Quantitative and qualitative alterations of acute-phase proteins in healthy elderly persons. *Age Ageing*, 1996 25,224-230.

Bartlett, GR. (1959). Phosphorus assay in column chromatography. *J Biol Chem*, 234, 466-468.

Benoit SC, Kemp CJ, Elias CF, Abplanalp W, Herman JP, Migrenne S, Lefevre AL, Cruciani-Guglielmacci C, Magnan C, Yu F, Niswender K, Irani BG, Holland WL, Clegg DJ. Palmitic acid mediates hypothalamic insulin resistance by altering PKC-theta subcellular localization in rodents. *J Clin Invest*. 2009 119, 2577-2589.

Blair, AS; Hajduch, E; Litherland, GJ; Hundal, HS. Regulation of glucose transport and glycogen synthesis in L6 muscle cells during oxidative stress: evidence for cross-talk between the insulin and SAPK2/p38 MAP kinase signaling pathways. *J Biol Chem*, 1999 274, 36293-36299.

Bligh, E. G. & Dyer, W. J. (1959). A rapid method of total lipid extraction and purification. *Can J Biochem Physiol*, 37, 911-917.

Boini, KM; Zhang, C; Xia, M; Poklis, JL; Li, P-L. Role of sphingolipid mediator ceramide in obesity and renal injury in mice fed a high fat diet. *J Pharmacol Exp Ther*. 2010, DOI:10.1124/jpet.110.168815.

Boyanovsky, B; Karakashian, A; King, K; Giltiay, N; Nikolova- Karakashian, M. Uptake and metabolism of low density lipoproteins with elevated ceramide content by human microvascular endothelial cells. *J Biol Chem*, 2003 278, 26992-26999.

Chavez, J. A. & Summers, S. A. (2003). Characterizing the effects of saturated fatty acids on insulin signaling and ceramide and diacylglycerol accumulation in 3T3-L1 adipocytes and C2C12 myotubes. *Arch Biochem Biophys*, 419, 101-109.

Chavez, JA; Knotts, TA; Wang, LP; Li, G; Dobrowsky, RT; Florant, GL; Summers, SA. A role for ceramide, but not diacylglycerol, in the antagonism of insulin signal transduction by saturated fatty acids. *J Biol Chem*, 2003 13, 10297-10303.

Cheng, Y; Ohlsson, L; Duan, R-D. Psyllium and fat in diets diferentially affect the activities and expressions of colonic sphingomyelinases and caspase in mice. *Br J Nutr*, 2004 91, 715-723.

Chocian, G; Chabowski, A; Zendzian-Piotrowska, M; Harasim, E; Łukaszuk, B; Górski, J. High fat diet induces ceramide and sphingomyelin formation in rat's liver nuclei. *Mol Cell Biochem.*, 2010 340, 125-131.

Costantini, C; Weindruch, R; Della Valle, G; Puglielli, L. A TrkA-to-p75^{NTR} molecular switch activates amyloid β-peptide generation during aging. *Biochem J*, 2005 391, 59-67.

Cutler, R. G. & Mattson, M. P. (2001). Sphingomyelin and ceramide as regulators of development and lifespan. *Mech Ageing Dev*, 122, 895-908.

Cutler, RG; Kelly, J; Storie, K; Pedersen, WA; Tammara, A; Hatanpaa, K; Troncoso, JC; Mattson, MP. Involvement of oxidative stress-induced abnormalities in ceramide and cholesterol metabolism in brain aging and Alzheimer's disease. *Proc Natl Aca. Sci USA*, 2004 101, 2070-2075.

de la Monte, SM. Insulin resistance and Alzheimer's disease. *BMB reports*, 2009 42, 475-481.

de Wilde, J; Mohren, R; van den Berg, S; Boekschoten, M; Dijk, KW; de Groot, P; Müller, M; Mariman, E; Smit, E. Short-term high-fat-feeding results in morphologival and metabolic adaptations in skeletal muscle of C57BL/6J mice. *Physiol Genomics*, 2008 32, 360-369.

Deevska, GM; Rozenova, KA; Giltiay, NV; Chambers, MA; White, J; Boyanovsky, BB; Wei, J; Daugherty, A; Smart, EJ; Reid, MB; Merrill, AH JR; Nikolova-Karakashian, M. Acid sphingomyelinase deficiency prevents diet-induced hepatic triacylglycerol accumulation and hyperglycemia in mice. *J Biol Chem*, 2009 284, 8359-8368.

Delgado, TC; Pinheiro, D; Caldeira, M; Castro, MM; Geraldes, CF; López-Larrubia P; Cerdán S; Jones JG. Sources of hepatic triglyceride accumulation during high-fat feeding in the healthy rats. *NMR Biomed*, 2009 22, 310-317.

Fox, TE; Houck, KL; O'Neill, SM; Nagarajan, M; Stover, TC; Pomianowski, PT; Unal, O; Yun, JK; Naides, SJ; Kester, M. Ceramide recruits and activates protein kinase C ζ (PKC ζ) within structured membrane microdomains. *J Biol Chem*, 2007 282, 12450-12457.

Granholm, AC; Bimonte-Nelson, HA; Moore, AB; Nelson, ME; Freeman, LR; Sambamurti, K. Effects of a saturated fat and high cholesterol diet on memory and hippocampal morphology in the middle-aged rat. *J Alzheimers Dis*, 2008 14, 133-145.

Guenther, GG; Peralta, ER; Rosales, KR; Wong, SY; Siskind, LJ; Edinger, AL. Ceramide starves cells to death by downregulating nutrient transporter proteins. *Proc Natl Acad Sci USA*, 2008 105, 17402-17407.

Gustavsson, L; Moehern, G; Torres-Morquez, ME; Benistant, C; Rubin, R; Hock, JB. The role of cytosolic Ca^{2+}, protein kinase C and protein kinase D in rat hepatocytes. *J Biol Chem*, 1994 269, 849-859.

Hajduch, E; Alessi, DR; Hemmings, BA; Hundal, HS. Constitutive activation of protein kinase Bα (PKBα) by membrane targeting promotes glucose and system A amino acid transport, protein synthesis and GSK3 inactivation in L6 muscle cells. *Diabetes*, 1998 47, 1006-1013.

Hajduch, E; Turban, S; Le Liepvre, X; Le Lay, S; Lipina, C; Dimopoulos, N; Dugail, I; Hundal, HS. Targeting of PKCζ and PKB to caveolin-enriched microdomains represents a crucial step underpinning the disruption in PKB-directed signaling by ceramide. *Biochem J*, 2008 410, 369-379.

He, X; Huang, Y; Li, B; Gong, CX; Schuchman, EH. Deregulation of sphingolipid metabolism in Alzheimer's disease. *Neurobiol Aging*, 2008 31, 398-408.

Holland, W. L. & Summers, S. A. (2008). Sphingolipids, insulin resistance, and metabolic disease: new insights from in vivo manipulation of sphingolipid metabolism. *Endocrine Reviews*, 29, 381-402.

Holland, WL; Brozinick JT; Wang, L-P; Hawkins, ED; Sargent, KM; Liu, Y; Narra, K; Hoehn, KL; Knotts, TA; Siesky, A; Nelson, DH; Karathanasis, SK; Fontenot, GK; Birnbaum, MJ; Summers, SA. Inhibition of ceramide synthesis ameliorates glucocorticoid-, saturated fat-, and obesity-induced insulin resistance. *Cell Metab*, 2007 5, 167-179.

Hudgins, LC; Hellerstein, MK; Seidman, C; Neese, R; Diakun, J; Hirsch, J. Human fatty acid synthesis is stimulated by a eucaloric low fat, high carbohydrate diet. *J Clin Invest*, 1996 97, 2081-2091.

JeBailey, L; Wanono, O; Niu, W; Roessler, J; Rudich, A; Klip, A. Ceramide- and oxidant-induced insulin resistance involve loss of insulin-dependent Rac-activation and actin remodeling in muscle cells. *Diabetes*, 2007 56, 394-403.

Lauter, C. J. & Trams, E. G. (1962). A spectrophometric determination of sphingosine. *J Lipid Res*, 3, 136-138.

Le Stunff, H; Giussani, P; Maceyka, M; Lépine, S; Milstien, S; Spiegel, S. Recycling of sphingosine is regulated by the concerted actions of sphingosine-1-phosphate phosphohydrolase 1 and sphingosine kinase 2. *J Biol Chem*, 2007 282, 34372- 34380.

Lin, J; Yang, R; Tarr, PT; Wu, PH; Handschin, C; Li, S; Yang, W; Pei, L; Uldry, M; Tontonoz, P; Newgard, CB; Spiegelman, BM. Hyperlipidemic effects of dietary saturated fats mediated through PGC-1 beta coactivation of SREBP. *Cell*, 2005 120, 261-273.

Long, S. D. & Pekala, P. H. (1996). Lipid mediators of insulin resistance: ceramide signaling down regulates GLUT4 gene transcription in 3T3-L1 adipocytes. *Biochem J*, 319, 179-184.

Lyn-Cook, LE, Jr; Lawton, M; Tong, M; Silbermann, E; Longato, L; Jiao, P; Mark, P; Wands, JR; Xu, H; de la Monte, S. Hepatic ceramide may mediate brain insulin resistance and neurodegeneration in type 2 diabetes and non-alcoholic steatohepatitis. *J Alzheimer Dis*, 2009 16, 715-729.

Marsh, J. B. & Weinstein, D. B. (1966). Simple charring method for determination of lipids. *J Lipid Res*, 7, 574-576.

Mattson, MP. Diet-brain connection: impact on memory, mood, aging and disease. Ed. Kluwer Academic Publishers; USA: Mattson, MP; 2002.

Merrill, AH, Jr; Lingrell, S; Wang, E; Nikolova-Karakashian, M; Vales, TR; Vance, DE. Sphingolipid biosynthesis de novo by rat hepatocytes in culture. *J Biol Chem*, *1995* 270, 13834–*13841*

Michalopoulos, G., & Pitot, H. C. (1975). Primary cultures of parenchymal liver cells on collagen membranes. *Exp Cell Res*, 33, 545-552.

Morino, K; Petersen, KF; Shulman, GI. Molecular mechanisms of insulin resistance in humans and their potential links with mitochondrial dysfunction. *Diabetes*, 2006 55, S9-S15.

Muskiet, FA; van Doormaal, IJ; Martini, IA; Wolthers, BG; van der Slick, W. Capillary gas chromatographic profiling of total long-chain fatty acids and cholesterol in biological materials. *J Chromatogr*, 1983 278, 231-244.

Nagle, CA; An, J; Shiota, M; Torres, TP; Cline, GW; Liu, Z-X; Wang, S; Catlin, RL; Shulman, GI; Newgard, CB; Coleman, RA. Hepatic overexpression of glycerol-sn-3-phosphate acyltransferase 1 in rats causes insulin resistance. *J Biol Chem*, 2007 282, 14807–14815.

Nagle, CA; Klett, EL; Coleman, RA. Hepatic triacylglycerol accumulation and insulin resistance. *J Lipid Res*, 2009 50, S74-S79.

Neschen, SK; Morino, LE; Hammond, D; Zhang, ZX; Liu, AJ; Romanelli, GW; Cline, RL; Pongratz, XM; Zhang, CS; Choi, CS; *Coleman, RA; Shulman, GI.* Prevention of hepatic steatosis and hepatic insulin resistance in mitochondrial acyl-CoA:glycerol-sn-3-phosphate acyltransferase 1 knock out mice. *Cell Metab*, 2005 2, 55–65.

Neumann, KF; Rojo, L; Navarrete, LP; Farías, G; Reyes, P; Maccioni, RB. Insulin resistance and Alzheimer's disease: molecular links & climical implications. *Curr Alzheimer Res*, 2008 5, 438-447.

Ogretmen, B; Pettus, BJ; Rossi, MJ; Wood, R; Usta, J; Szulc, Z; Bielawska, A; Obeid, LM; Hannun, YA. Biochemical mechanisms of the generation of endogenous long chain ceramide in response to exogenous shirt chain ceramide in the A549 human lung adenocarcinoma cell line. Role for endogenous ceramide in mediating the action of exogenous ceramide. *J Biol Chem*, 2002 277, 12960-12969.

Parrott, M. D. & Greenwood, C. E. (2007). Dietary influences on cognitive function with aging: from high-fat diets to healthful eating. *Ann N Y Acad Sci*, 1114, 389-397.

Patil, S. & Chan, C. (2005). Palmitic and stearic acids induce Alzheimer-like hyperphosphorylation of tau in primary rat cortical neurons. *Neurosci Lett*, 384, 288-293.

Patil, S; Balu, D; Melrose, J; Chan, C. Brain region-specificity of palmitic acid-induced abnormalities associated with Alzheimer's disease. *BMC Research Notes*, 2008 1, 20.

Patil, S; Melrose, J; Chan, C. Involvement of astroglial ceramide in palmitic acid-induced Alzheimer-like changes in primary neurons. *Eur J Neurosci*, 2007 26, 2131-2141.

Patil, S; Sheng,, L; Masserang, A; Chan, C. Palmitic acid-treated astrocytes induce BACE1 upregulation and accumulation of C-terminal fragment of APP in primary cortical neurons. *Neurosci Lett*, 2006 406, 55-59.

Puglielli, L; Tanzi, RE; Kovacs, DM. Alzheimer's disease: the cholesterol connection. *Nat Neurosci*, 2003 6, 345-351.

Rutkute, K; Asmis, RH; Nikolova-Karakashian, MN. Regulation of neutral aphingomyelinase-2 by GSH: a new insight to the role of oxidative stress in aging-associated inflammation. *J Lipid Res*, 2007 48, 2443-2452.

Salinas, M; Lopez-Valdaliso, R; Martin, D; Alvarez, A; Cuadrado, A. Inhibition of PKB/Akt by C2-ceramide involves activation of ceramide-activated protein phosphatase in PC12 cells. *Mol Cell Neurosci*, 2000 15, 156-169.

Sampath, H; Miyazaki, M; Dobrzyn, A; Ntambi, JM. Srearoyl-CoA desaturase-1 mediates the pro-lipogenic effects of dietary saturated fat. *J Biol Chem*, 2007 282, 2483-2493.

Samuel, VT; Liu, Z-X; Qu, X; Elder, BD; Bilz, S; Befroy, D; Romanelli, AJ; Shulman, GI. Mechanism of hepatic insulin resistance in non-alcoholic fatty liver disease. *J Biol Chem*, 2004 279, 32345- 32353.

Sathishkumar, S; Boyanovsky, B; Karakashian, AA; Rozenova, K; Giltiay, NV; Kudrimoti, M; Mohiuddin, M; Ahmed, MM; Nikolova-Karakashian, M. Elevated sphingomyelinase activity and ceramide concentration in serum of patients undergoing high dose spatially fractionated radiation treatment: implications for endothelial apoptosis. *Cancer Biol Ther*, 2005 4, 979-986.

Schmitz-Peiffer, C. & Biden, T.J. (2008). Protein kinase C function in muscle, liver, and ß-cells and its therapeutic implications for type 2 diabetes. *Diabetes*, 57, 1774-1783.

Schmitz-Peiffer, C; Craig, DL; Bidn, TJ. Ceramide generation is sufficient to account for the inhibition of the insulin-stimulated PKB pathway in C2C12 skeletal muscle cells pretreated with palmitate. *J Biol Chem*, 1999 274, 24202-24210.

Schwarz, JM; Neese, RA; Turner, S; Dare, D; Hellerstein, MK. Short-term alterations in carbohydrate energy intake in humans. Striking effects on hepatic glucose production, de novo lipogenesis, lipolysis, and whole-body fuel selection. *J Clin Invest*, 1995 96, 2735-2743.

Shah, C; Yang, G; Lee, I; Bielawski, J; Hannun, YA; Samad, F. Protection from high fat diet-induced increase in ceramide in mice lacking plasminogen activator inhibitor 1. *J Biol Chem*, 2008 283, 13538-13548.

Solfrizzi, V; D'Introno, A; Colacicco, AM; Capurso, C; Del Parigi, A; Capurso, S; Gadaleta, A; Capurso, A; Panza, F. Dietary fatty acids intake: possible role in cognitive decline and dementia. *Exp Gerontol*, 2005 40, 257-270.

Stratford, S; DeWald, DB; Sammers, SA. Ceramide dissociates 3′-phosphoinositide production from pleckstrin homology domain translocation. *Biochem J*, 2001 354, 359-368.

Teruel, T; Hernandez, R; Lorenzo, M. Ceramide mediates insulin resistance by tumor necrosis factor-α in brown adipocytes by maintaining Akt in an inactive dephosphorylated state. *Diabetes*, 2001 50, 2563-2571.

Uauy, R. & Dangour, A. (2006). Dietary lipids and the brain during development and ageing. *Scand J Food Nutr*, 50, 27-32.

Ugi, S; Imamura, T; Maegawa, H; Egawa, K; Yoshizaki, T; Shi, K; Obata, T; Ebina, Y; Kashiwagi, A; Olefsky, JM. Protein phosphatase 2A negatively regulates insulin's metabolic signaling pathway by inhibiting Akt (protein kinase B) activity in 3T3-L1 adipocytes. *Mol Cell Biol*, 2004 24, 8778-8789.

Vranckx, R; Savu, L; Lambert, N; de Conchard GV; Grosse, R; Mourey, MS; Corman, B. Plasma proteins as biomarkers of the aging process. *Am J Physiol*, 1995 268, R536-548.

Winocur, G. & Greenwood, C. E. (2005). Studies of the effects of high fat diets on cognitive function in a rat model. *Neurobiol Aging*, 26 Sappl 1, 46-49.

Zinda, MJ; Vlahos, CJ; Lai, MT. Ceramide induces the dephosphorylation and inhibition of constitutively activated Akt in PTEN negative U87mg cells. *Biochem Biophys Res Commun*, 2001 280, 1107-1115.

In: Saturated Fats
Editor: Joseph P. Langella
ISBN: 978-1-61209-566-0

Chapter IV

Insect Lipids Mobilized by Adipokinetic Hormones

***Dalibor Kodrík*[1,2], *Aleš Tomčala*[3], *Iva Bártů*[1,2] *and Radomír Socha*[1]**

[1]Institute of Entomology, Biology Centre, Academy of Sciences, České Budějovice, Czech Republic
[2]Faculty of Science, University of South Bohemia, České Budějovice, Czech Republic
[3]Institute of Organic Chemistry and Biochemistry, Academy of Sciences, Praha, Czech Republic

Abstract

A major transport form of lipids in insect haemolymph are diacyglycerols (DGs), predominantly bound to carrier protein lipophorins, in contrast to vertebrates where the lipid transport is established by free fatty acids (FAs). Mobilization of the DGs from the fat body and their transport to the target effectors, mostly to working muscle cells, is under the control of adipokinetic hormones (AKHs). The AKHs play a crucial role in the control of insect energy metabolism - they behave as typical stress hormones by stimulating catabolic reactions, making energy more available, while inhibiting synthetic reactions. The DG molecular species and their FA composition were investigated by electrospray mass spectrometry (ESI-MS) in haemolymph of *Locusta migratoria* and *Pyrrhocoris apterus* after the application of their innate AKHs (Locmi-AKH-I, AKH-II and AKH-III in *L. migratoria,* and Pyrap-AKH-I and Peram-CAH-II in *P. apterus*). Mobilization of DGs from the fat body after injection of AKH was not uniform. AKHs mobilized the DGs selectively with preference of those possessing the C18 and C16 FAs, especially 18:1 in *L. migratoria* and 18:2 in *P. apterus.* The fat body FAs with a carbon chain longer than 18 did not participate in the mobilization. In *L. migratoria,* Locmi-AKH-I preferred release of DGs with unsaturated FAs, while AKH-II and AKH-III stimulated DG release with saturated FAs. In *P. apterus,* significant quantitative differences between Pyrap-AKH-I and Peram-CAH-II were also shown: the former preferred linoleic acid (18:2), the later oleic acid (18:1). Heterologous assays, where the AKHs from one species were applied to another one, suggested that the activation of DGs and/or FAs was rather species- than hormone-specific. The relationship between the AKH action and FA distribution in DGs are summarized and discussed in detail.

Introduction

In addition to various necessary nutritive compounds, most insects have dietary needs for lipids and fatty acids (FA). Many studies have demonstrated that triacylglycerols (TGs) are a major lipid class serving as a reserve of metabolic energy, which can be mobilized and utilized in response to bioenergetic demands. In insects, TGs occur predominantly in the fat body (FB), an organ analogous to vertebrate adipose tissue and liver, where they represent about 90% of the stored lipids (Arrese & Wells, 1994; Canavoso et al., 1998). Some TGs are also present in the haemolymph. However, diacylglycerols (DGs), predominantly bound to carrier protein lipophorins, represent the major haemolymph lipids (Chino & Downer, 1982). At least 90% of haemolymph DGs are present as the *sn*-1,2-isomers and serve as the transport lipid form in insects, in contrast to vertebrates where the lipid transport is established by free fatty acids (FAs) (Van der Horst, 1982; Arrese et al., 1996).

Insects digest and absorb lipids analogous to vertebrates, but with some important differences. Contrary to vertebrates, the insect system is more efficient because lipids are delivered to tissues, for the most part, without internalization and destruction of lipophorin (Soulages & Wells, 1994; Aresse et al., 2001). Insects do not have bile salts, but have developed alternative strategies to facilitate lipid solubility: they use lumenal glycolipids, form fatty acyl-amino acid complexes, and FAs and lysophospholipid micelles (Turunen & Crailsheim, 1996). The midgut cells produce lipases that hydrolyze the dietary TGs, forming monoacylglycerols and free FAs (Hoffman & Downer, 1979). The absorbed FAs are converted into DGs using the phosphatidic acid pathway (Canavoso & Wells, 2000). The DGs are rapidly converted into TGs, which serve as a reservoir for absorbed FAs, or are exported directly to the FB via haemolymph (Canavoso & Wells, 2000) to form the FB TGs.

The DG molecular species are characterized by individual FAs. Several dozen FAs have been described in the insect haemolymph, however, the major proportion is represented just by 8-9 FAs possessing 12-18 carbon atoms and including saturated FAs: lauric (12:0), myristic (14:0), palmitic (16:0) and stearic (18:0), the monounsaturated FAs: myristoleic (14:1), palmitoleic (16:1) and oleic (18:1) and polyunsaturated FAs: linoleic (18:2) and linolenic (18:3). The monounsaturated FAs are composed primarily of the cis-Δ-9-isomer and the C18 polyunsaturated FAs are the cis-Δ-9-12- and cis-Δ-9-12-15 isomers (Downer, 1985; Schneider & Dorn, 1994).

Metabolism of lipids, especially their mobilization from FB stores into the haemolymph, is hormonally controlled. The principal role is played by a group of neuropeptides called adipokinetic hormones (AKHs). Up to now, almost 50 insect AKHs have been isolated and characterized from representatives of many insect orders (Gäde et al., 1997; Gäde, 2009). Together with one crustacean represent-tative that was found recently also in bugs (Heteroptera) (Gäde et al., 2003; Kodrík et al., 2010), a chromatophorotropin - red pigment concentrating hormone (RPCH) (Fernlund & Josefsson, 1972), they form an AKH/RPCH peptide family. The AKHs are synthesized, stored and released by neurosecretory cells of the corpora cardiaca - an endocrine gland linked with the insect brain. With some simplification, an analogy can be drawn between the brain (neurosecretory cells) and corpora cardiaca secretory system in insects, and the hypothalamus and the pituitary gland system (adenohypophysis and neurohypophysis) in vertebrates (Scharrer & Scharrer, 1944).

Biochemical and molecular studies revealed that AKHs are short peptides (mostly 8 - 10 amino acid residues) with both termini blocked (see review Gäde et al., 1997), except for an unusual AKH of the butterfly *Vanessa cardui* (Köllisch et al., 2000) that is an 11-mer in which the AKH is not processed completely from the prohormone and occurs in a C-terminally extended form. Recently, a phosphorylated member of the AKH/RPCH family was identified in the beetle *Trichostetha fascicularis* (Gäde et al., 2006), which represents the first phosphorylated invertebrate neuropeptide. For a long time it was supposed that a major function of AKHs is the control of insect metabolism. However, these peptides are pleiotropic, with a number of actions attached to their metabolic role (Gäde et al., 1997; Gäde & Marco, 2006; Kodrík, 2008). Generally, they behave as typical stress hormones by stimulating catabolic reactions (mobilize lipids, carbohydrates and/or certain amino acids), making energy more available, while inhibiting synthetic reactions (synthesis of lipids, RNA and proteins) that are momentarily less important. The stress metabolic action of AKHs can be supplemented by other processes comprising of the increase of muscle tonus (O'Shea et al., 1984), stimulation of heart beat (Scarborough et al., 1984) and general locomotion (Socha et al., 1999; Kodrík et al., 2000), enhancement of immune response (Goldsworthy et al., 2002, 2003), inhibition of egg maturation (Lorenz, 2003) and/or activation of antioxidant mechanisms (Kodrík et al., 2007).

Insect fat body is the main target tissue for AKH action. Signal transduction at the cellular level is well-documented there (see Van der Horst et al., 2001; Gäde & Auerswald, 2003). Because AKHs as peptidic hormones are not able to penetrate freely the cell membrane, their message is transduced via specific membrane receptors (Park et al., 2002; Staubli et al., 2002; Hansen et al., 2006; Kaufmann & Brown, 2006). The receptor is linked to a G-protein that activates two (or three) biochemical pathways leading to the production of energy-rich substrates like DGs, trehalose or proline. These compounds are transported to the target effectors, mostly to working muscle cells (for details see Gäde & Auerswald, 2002, 2003). In the case of the lipid mobilization, the AKHs stimulate lipases to break down TGs to monoacylglycerols, which are afterwards re-acylated to DGs. The DGs are then released from the FB into the haemolymph. The cAMP, Ca^{2+} and protein kinases are apparently involved in the process. In the haemolymph, the AKHs play a role in a rearrangement of existing lipoproteins and in the mobilization of stored fuels during episodes of flight or other types of insect locomotion (e.g. enhanced walking, swimming etc.).

It is interesting to note that some insects have more than one (2 or 3) native AKH peptide. The best-known example is the African migratory locust *Locusta migratoria*, a classical insect model for AKH research that possesses three different AKHs: Locmi-AKH-I, AKH-II and AKH-III (Goldsworthy, 1994). Reasons for the higher number of the hormones are not completely clear: is it only an evolutionary accident or does it have specific metabolic importance and meaning? In the present study, this question was investigated rightly using the *L. migratoria,* and also a representative of Heteroptera, the firebug *Pyrrhocoris apterus*. Two AKHs were characterized in this bug - Pyrap-AKH (Kodrík et al., 2000) and Peram-CAH-II (Kodrík et al., 2002), so both insect species are convenient models for the study (for other reasons see also below).

Most of the research on the metabolic function of AKHs in insects has been done using a classical approach, i.e. by means of conspecific tests, in which each studied insect species was treated with its (their) native specific AKH(s). It is presumed that many peptide hormones have been conserved during evolution with function being less conserved than structure (Ebberink et al., 1989; Satake et al., 1999). However, it cannot be excluded that in the case of AKHs, the type of mobilized substrates used as fuels for insect locomotion (either flight or enhanced walking activity) can also be conservative due to evolutionarily pressures on a particular insect species. Therefore, an important question arises, if the metabolic effects of AKHs depend on the recipient species or rather on the structure of the AKH tested. In other words, whether an activation of DGs and/or FAs is rather species- than hormone-specific. To solve this question, heterologous assays, where inter-species applications of AKHs form *L. migratoria* on *P. apterus*, and vice versa, were performed.

The main aim of the present study was consequently to find (1) whether the DGs and FAs in *L. migratoria* and *P. apterus* are selectivity mobilized by their specific native AKHs, and (2) whether the activation of DGs and/or FAs is species- or hormone-specific. The results are supposed to contribute to a better understanding of AKH-controlled insect lipid metabolism, and to an extension of our knowledge why certain insect species possess more than one AKH.

Lipid Mobilization in *Locusta Migratoria*

A typical representative of the flying insect species is *L. migratoria,* in which the AKHs predominantly control the mobilization of lipids, particularly during long-distance migratory flights. Nevertheless, this species is a classical model for the AKH studies being used from the 1970's. A significant part of the AKH functions in metabolism and other attached processes mostly related to a stress on molecular, biochemical and physiological levels has been revealed on this species (see Goldsworthy, 1990; Gäde et al., 1997; Van der Horst et al., 2001; Gäde & Goldsworthy, 2003; Kodrík, 2008). *L. migratoria* possesses three AKHs that are synthesized in the glandular lobes of the corpora cardiaca: the decapeptide Locmi-AKH-I (pGlu-Leu-Asn-Phe-Thr-Pro-Asn-Trp-Gly-Thr-NH2; Stone et al., 1976) and two octapeptides - Locmi-AKH-II (pGlu-Leu-Asn-Phe-Ser-Ala-Gly-Trp-NH2; Siegert et al., 1985) and Locmi-AKH-III (pGlu-Leu-Asn-Phe-Thr-Pro-Trp-Trp-NH2; Oudejans et al., 1991). Injection of the AKHs into a locust body significantly increases the level of DGs in the haemolymph (Beenakkers et al., 1985; Goldsworthy et al., 1997). It is supposed that most of them are revealed as *sn*-1,2-DGs, which is typical not only for *L. migratoria* (Beenakkers et al., 1985), but was also proven in other insect species: for example in *Manduca sexta* haemolymph the *sn*-1,2-DG isomers represented 88.4% of total DG content, while the *sn*-2,3- and 1,3-isomers represented only 4.6% and 7%, respectively (Arrese & Wells, 1997). A relative distribution of the DG molecular species after the injection of 10 pmol Locmi-AKH-I, AKH-II and AKH-III into the adult males and females has been studied recently (Tomčala et al., 2010). The ESI-MS^2 analysis revealed that various DGs were mobilized selectively after the hormonal treatment (Figure 1) with maximal stimulation of 594 Da DG (for all Locmi-AKHs) that contains palmitic and oleic acids in its molecule i.e. the DG 16:0/18:1. The data was calculated directly from ESI-MS^2 measurements, where mutual subtraction of total DG levels in resting (time 0) and treated (90 min later) locusts were considered to be 100%. Injection of the solvent only (control treatment) had no significant effect (data not shown) on total lipid elevation. The one-way ANOVA analysis of the differences among the AKHs indicated that Locmi-AKH-I significantly preferred mobilization of DGs mw 596 Da (FAs - 16:0, 18:0), 612 Da (FAs - 18:2, 18:3, 18:4), 614 Da (FAs - 18:2, 18:3) and 618 Da (FAs - 18:0, 18:1, 18:2, 18:3), while AKH-II and -III preferred DGs mw 566 Da (FAs - 14:0, 16:0, 16:1, 18:1), 568 Da (FAs - 14:0, 16:0, 18:0) and 622 Da (FAs - 18:0, 18:1). No significant difference between Locmi-AKH-II and AKH-III was recorded.

The FA analyses of the DGs mobilized after the injections of Locmi-AKHs, and those of the TGs from intact FB revealed that the AKH-induced selectivity is evident also on the FA level (Figure 2). Comparison of the two FA pools showed that (1) there is a different spectrum of the FAs mobilized by AKHs into the haemolymph, and the FA pool in the FB, (2) that the C16 and C18 FAs play a crucial role, and (3) that significant changes in the levels were recorded for FAs 14:0, 16:0, 18:0, 18:1, 18:2 and 18:3. All three Locmi-AKHs preferably mobilized DGs with oleic (18:1) and linolenic acids (18:3); AKH-I additionally mobilized linoleic acid (18:2) containing DGs, while AKH-II and AKH-III mobilized stearic acid (18:0) containing DGs. Relative proportion of palmitic acid (16:0) (and also myristic acid 14:0 - but its level was very low) in Locmi-AKHs treated haemolymph was surprisingly lower than in the FB.

The differences between the AKH specificity might be explained by the putative existence of two different acylglycerol pools in the FB and their complicated reconstruction to the DGs (for details, see Beenakkers et al., 1985). It is interesting to note that almost exclusively the C18 and C16 FAs are involved in AKH elicited lipid mobilization. This result supports the locust flying studies, which revealed that the C16 and C18 (C34 and C36 DGs) were predominantly mobilized FAs. They included mainly oleic and palmitic acids, and their relative composition remained fairly constant during flight periods of up to 6 hours

(Beenakkers et al., 1985). This data is not perhaps so surprising, as eukaryotic FA-synthases synthesize mainly C16 saturated, and also C18 and C14 FAs. The specificities of the chain-elongating and chain-terminating enzymes complement each other perfectly, ensuring that these FAs are the major products (Rangan & Smith, 2002).

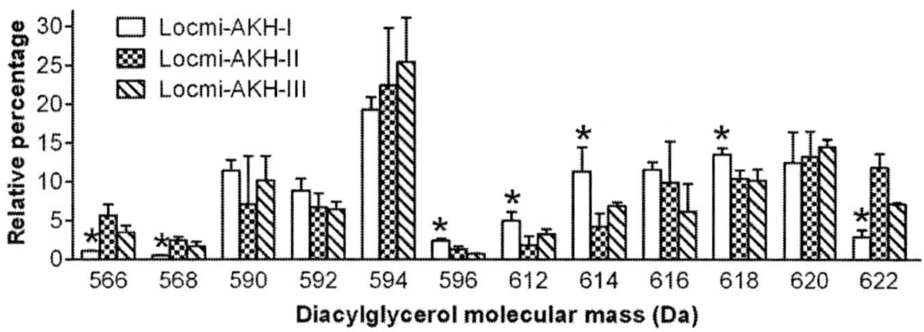

Figure 1. Relative percentage of the individual DG species participating in the total elevation of haemolymph lipids (=100 %) of *L. migratoria* after the treatment by Locmi-AKHs. The effect of the injection of a solvent only was negligible (data not shown). Statistically significant differences at the 5% level evaluated by one-way ANOVA (AKH-I vs. AKH-II, which is identical to AKH-I vs. AKH-III; no significant differences between AKH-II vs. AKH-III were recorded) are indicated by asterisks. The bars represent SDs, n=3. For details, see also the corresponding text (modified according to Tomčala et al., 2010).

All three Locmi-AKHs preferably mobilize the DGs with the presence of oleic acid (18:1); thus oleic acid is responsible for 35 - 40 % of the whole lipid mobilization. Prominent position of this FA reflects its important role in covering of insect energy demands. The percentage of oleic acid in the FB lipids is generally higher in a potentially migrating gregarious phase than in the solitary one of *S. gregaria*. The DGs containing oleic acid plus unsaturated C18 linoleic and linolenic acids exhibited a significant haemolymph increase during the flight of this species. They significantly fluctuated not only in relation to phase, but also to their age (Schneider & Dorn, 1994). High content of oleic acid was recorded also in FBs of several more insect species including the beetle *Zophobas atratus* (Howard & Stanley-Samuelson, 1996). Oleic acid is also the most common DG-lipophorin FA in the blood-sucking bug *Panstrongylus megistus* (Canavoso et al., 2004).

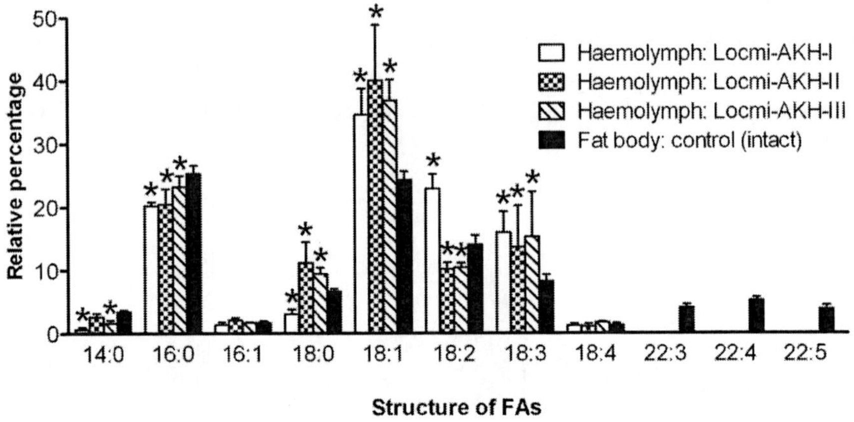

Figure 2. Relative percentage of the FAs contained in the haemolymph DGs mobilized by Locmi-AKHs injected into the *L. migratoria*, and their comparison with FAs in the intact FB TGs. Statistically significant differences at the 5% level evaluated by t-test (FAs in haemolymph DGs vs. FAs in FB TGs) are indicated by asterisks. The showed TG FAs represent 98% of the total FAs (other detected FAs were present in negligible amounts). The bars represent SDs, n=3 for haemolymph samples and n=10 for FB samples (according to Tomčala et al., 2010).

Figure 2 makes also possible the evaluation of differences in mobilization of the particular FAs among the AKHs. The one-way ANOVA analysis (5 % significance level, statistical data not shown in Figure 2) of the differences indicated that, in spite of some overlapping, Locmi-AKH-I preferred unsaturated, while AKH-II and AKH-II preferred saturated FAs. It was shown that linoleic acid (18:2) was significantly, primarily mobilized by AKH-I, while stearic (18:0) and myristic (14:0) FAs were significantly preferred by AKH-II and AKH-III. Within the rest of tested FAs (16:0, 16:1, 18:1, 18:3, 18:4) there were no significant differences among the hormones recorded. This accords with the Murata and Tojo (2002) finding that in locusts, unsaturated FAs in TGs are mainly used as an energy source during flight, and that Locmi-AKH-I is more involved in lipid mobilization during flight than Locmi-AKH-II and -III (Goldsworthy, 1994).

It seems that FAs with a carbon chain longer than 18 play no role in the AKH elicited lipid mobilization, although they are present in the FB (Figure 2). A question erases what is their possible role when they are not involved in the energy metabolism. There is no direct data about a role of those FAs in locust FB available, but they might have structural (cellular membranes) or regulatory (arachidonic acid) functions. The absence of FAs with C>18 in DGs released into the haemolymph may also be explained by their possible position at *sn*-3, which stereospecific configuration is not commonly mobilized into the haemolymph (Arrese & Wells, 1997; Van der Horst, 1982).

A question also arises if the above mentioned specialization of particular Locmi-AKHs could contribute to the explanation of the fact why some insect species possess more than one AKH - a phenomenon that is not satisfactory elucidated. A certain degree of AKH specialization was recorded a long time ago. The Locmi-AKH-I is more potent in a lipid mobilization assay than the Locmi-AKH-II and AKH-III (Goldsworthy, 1994). Furthermore, the Locmi-AKH-II is somewhat stronger than AKH-I in activation of glycogen phosphorylase (Oudejans et al., 1992; Van der Horst et al., 2001). Also, the inhibition of total RNA synthesis in *L. migratoria* FB is decreasing in the order Locmi-AKH-III > AKH-II > AKH-I (Kodrík & Goldsworthy, 1995). An appreciable degree of AKH selectivity in mobilization of DGs, and hence, FAs, has been demonstrated in this chapter. All this data is promising, but the phenomenon is far from understood.

Lipid Mobilization in *Pyrrhocoris Apterus*

Unlike the flying *L. migratoria*, the firebug *P. apterus* (Heteroptera) is a representative of the non-flying insect species using walking as a way of locomotion, however, similarly as the locust burning predominantly lipids to fuel its energy demands (Martin, 1969a,b; Socha & Kodrík, 1999). *P. apterus* produces two AKH peptides: Pyrap-AKH (pGlu-Leu-Asn-Phe-Thr-Pro-Asn-Trp-NH_2; Kodrík et al., 2000) and Peram-CAH-II (pGlu-Leu-Thr-Phe-Thr-Pro-Asn-Trp-NH_2; Kodrík et al., 2002). Both are synthesized and released into the haemolymph from the corpora cardiaca as common in other insect species. There is a lot of information available about a role of AKHs in metabolism, locomotion and in a lot of various stressing situations of this bug summarized by Kodrík (2008). The effect of Pyrap-AKH and Peram-CAH-II on a relative distribution of DG molecular species has been recently studied by Bártů et al. (2010).

P. apterus is characterized by a conspicuous, non-functional wing polymorphism and produces the macropterous (long-winged) and brachypterous (short winged) morphs (Socha, 1993). Both morphs are flightless (Socha & Zemek, 2000) and differ in their life history strategies (Socha & Zemek, 2003; Socha et al., 2005; Socha & Šula, 2008). Only macropterous females have been used for the experiments presented in this chapter (for reasons, see Socha & Kodrík, 1999). The ESI-MS^2 analysis of the DG molecular species after the injection of 10 pmol Pyrap-AKH and Peram-CAH-II revealed a presence of a relatively large number of representatives, of which about 10 were present only in negligible amounts and had no substantial effect on the total lipid level in the haemolymph. A characteristic set of the next 10 DGs that made up more than 90 % of the total DG in the haemolymph were selected for further study. Their molecular masses ranged from 590 to 622 Da. As in the experiments with *L. migratoria,* it was supposed that most of the DGs contain acyls in the *sn*-1,2-DG position (see previous paragraph).

The percentage of individual DGs presented in the haemolymph of the bugs after the Pyrap-AKH or the Peram-CAH-II injection is shown in Figure 3. The results were calculated from the ESI-MS^2 analyses and as mentioned above, the subtraction of the total DG level in resting (time 0) and treated (90 min later) bugs were considered to be 100%. The results revealed that just four or five DGs - 592 Da (16:0, 18:2), 616 Da (18:1, 18:2, 18:3), 618 Da (18:0, 18:1, 18:2, 18:3), 620 Da (18:0, 18:1) and less so 594 Da (16:0, 18:1) - were responsible for about 90% of the increase in lipids induced by the hormonal injection; as evident, these DGs contained C18 and C16 FAs only. A common t-test showed some statistical differences in mobilized DGs between the AKHs: Pyrap-AKH significantly mobilized more 616 Da DG and Peram-CAH-II more of the 594 and 620 Da DGs.

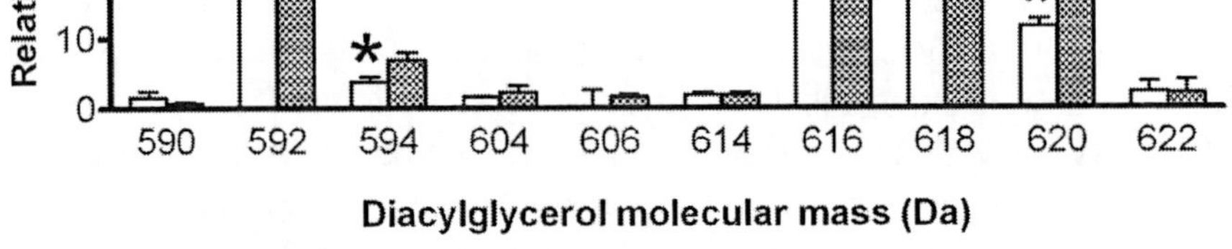

Figure 3. Relative percentage of the individual DG species participating in the total elevation of haemolymph lipids (=100 %) of *P. apterus* after the treatment by Pyrap-AKH and Peram-CAH-II. The effect of the injection of solvent only was negligible (data not shown). Statistically significant differences at the 5% level evaluated by t-tests (DG mobilized by Pyrap-AKH vs. DG mobilized by Peram-CAH-II) are indicated by asterisks. The bars represent SD, n=3. For details see also the corresponding text (according to Bártů et al., 2010).

The HPLC/ESI-MS analysis of TGs from FB of *P. apterus* resulted in the characterization of 21 different molecular species mostly presented in low amounts (data not shown). Only 3 TGs of mws 878 Da (18:1, 18:2, 18:3), 880 Da

(18:1, 18:2) and 882 Da (18:0, 18:1, 18:2) each make up more than 10% of the total the TG content of FB. It is interesting that all of these three TGs contained just C18 (mostly) unsaturated FAs. FB DGs made up only 2 % of the FB lipids and were omitted.

The FA analyses of the DGs mobilized after the injections of Pyrap-AKH and Peram-CAH-II, and the TGs from intact FB revealed similarities with the *L. migratoria* results, but also some interesting differences. The results presented in Figure 4 indicate that (1) as mentioned above, mainly the C18 and partially C16 FAs have a crucial role - FAs with a carbon chain longer than 18 and shorter than 16 do not have a role in lipid mobilization, although present in FBs; (2) percentual pattern of the AKH mobilized FAs in the haemolymph does not reflect that in the FB, which indicates that particular FAs are preferentially mobilized; (3) the differences in FA levels in hormonally mobilized DGs and those in the FB TGs were recorded for palmitic (16:0), stearic (18:0), oleic (18:1), linoleic (18:2) and linolenic (18:3) acids, whereas the levels of palmitic and linoleic acids were higher, and those of stearic, oleic and linolenic acids were lower in haemolymph of hormonally treated bugs than in the FBs of the control bugs. It is evident that there is also a selectivity of AKH-mobilized FAs in *P. apterus*, as well as it was proven above for *L. migratoria* (Figure 2). On the other hand, the spectrum of preferred FAs between the species differs. The reasons for this phenomenon can include differences in the species biology, food or living conditions, or some biochemical aspects such as a difference in the composition of the AKHs.

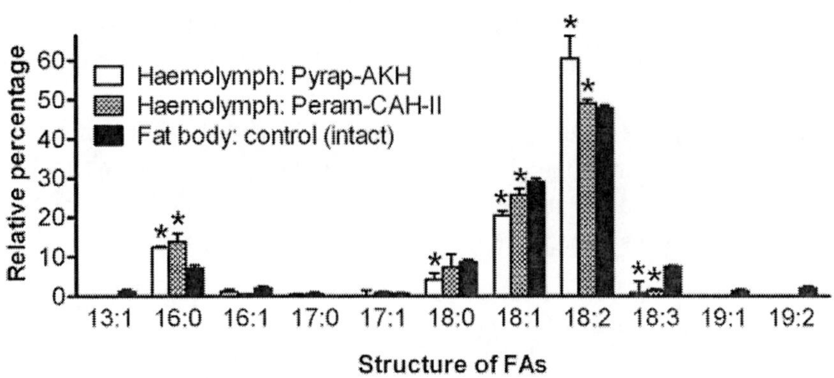

Figure 4. Relative percentage of the FAs contained in the haemolymph DGs mobilized by Pyrap-AKH and Peram-CAH-II injected into the *P. apterus*, and their comparison with FAs in the intact FB TGs. Statistically significant differences at the 5% level evaluated by t-tests (FAs in haemolymph DGs vs. FAs in FB TGs) are indicated by asterisks. The TG FAs make up 100% of the total FAs (including those present in negligible amounts). The bars represent SD, n=3 for haemolymph samples and n=12 for FB samples (according to Bártů et al., 2010).

Statistical evaluations (t-test, 5 % significance level) of the effect of individual hormones on the mobilization of particular FAs (statistical data not shown in Figure 4) indicated there was a significant difference between the effect of injected AKHs: the Pyrap-AKH mobilized more linoleic acid (18:2) while the Peram-CAH-II more oleic acid (18:1). Within the rest of tested FAs (16:0, 16:1, 17:0, 17:1, 18:0, 18:3), there were no significant differences between the hormones recorded. This result seems to be more complicated than in locusts, where at least partial preferential mobilization of unsaturated (Locmi-AKH-I) and saturated (Locmi-AKH-II and AKH-III) was recorded (Figure 2). There is no physiological background explaining the obtained data in *P. apterus* available.

Figure 4 also shows that linoleic acid (18:2) was the most abundant FA involved in the AKH induced mobilization of DGs, making up 60% of the lipids mobilized by Pyrap-AKH and 50% of that by Peram-CAH-II. Nevertheless, the linoleic acid content of the FB is around 45%, which indicates the importance of this FA in energy metabolism and the specificity of the Pyrap-AKH in its mobilization. The importance of linoleic acid in *P. apterus* was recorded a long time ago by Martin (1969a) who found that it made up 55 - 60% of the lipids in the FB and fluctuated around 50 - 55% in the ovaries at the beginning of the adult development. Those results suggest that linoleic acid might play a significant role in *P. apterus* in the most energetic demanding processes in insects controlled by AKH: locomotion and vitellogenin synthesis. In other insects, at least where the data is available, oleic acid (18:1) seems to have this role. This FA is responsible for 35 - 40% of the AKH induced lipid mobilization in *L. migratoria* (Figure 2), and its content in the FB is about 25%, which is identical to that of another frequent FA, palmitic acid (16:0) in this species (Figure 2, Beenakkers & Scheres, 1971). Nevertheless, oleic acid is the second most abundant FA in *P. apterus* (Figure 4) reaching about 20 - 25% in mobilized DGs and about 30% in FB TGs. The fact that its content in the FB is higher than in the AKH treated haemolymph suggests that oleic acid has also other functions. However, in the absence of direct information, it remains a speculation.

Heterologous Assays

The results of above mentioned experiments on *L. migratoria* and *P. apterus* demonstrate that mobilization of DGs and/or FAs from FBs after the AKH injection is not uniform and that the lipid molecular species are mobilized selectively. A question arises if those characteristics are species- or rather hormone-specific - in other words, if the selectivity depends on a character of target tissues including their receptors, or on a composition of the used hormones. The question can be simply solved by a heterologous bioassay. This approach is commonly used in the AKH research (Gäde et al., 1997; Socha et al., 2004) and includes a test of an AKH from donor insect species on a biological activity of an acceptor. The heterologous bioassay was used in the experiments when 10 pmol of Locmi-AKHs were injected into the *P. apterus* bodies, and reciprocally, 10 pmol of Pyrap-AKH or Peram-CAH-II into the *L. migratoria* ones. The results of ESI-MS^2 analysis revealed that the injection of *P. apterus* AKHs into the locust body (Figure 5A) mobilized predominantly the DG of mw 594 Da, and in a lesser degree, the DGs 590, 592, 616, 618, 620 and 622 Da. No statistically significant differences between the hormones (t-test) were recorded. The DG mobilized pattern is similar to that found in Figure 1 using the *L. migratoria* AKHs. This suggests that the pattern of lipid mobilization is species-specific and that the target tissue - mostly FB - and a quality of its TGs are generally responsible for the spectrum of mobilized DGs. The conclusion was verified on the *P. apterus* model. The Locmi-AKHs injections into the bugs bodies (Figure 5B) showed the preferential mobilization of the DGs with mws 592, 616, 618 and 620 Da and in less degree, 594 Da - the pattern that can be seen in the bugs injected by their own AKHs (Figure 3). Again, no statistically significant difference between the hormones (one-way ANOVA, Tukey post test) was recorded.

The FA analyses of both the mobilized DGs and FB TGs showed that the species-specific pattern is also evident on the FA level (Figure 6). The locust FB released predominantly oleic acid (18:1) (Figure 6A) being treated by the Pyrap-AKH and Peram-CAH-II. Similarly, as is shown in Figure 2 (conspecific test), the other FAs follow in decreasing order: 16:0 > 18:2 > 18:3 > 18:0. The treatment by both AKHs was responsible for a significantly higher level (t-test) of palmitoleic acid (16:1), stearic acid (18:0), oleic acid (18:1) and linolenic acid (18:3) in haemolymph compared to their level in the FB. The levels of myristic acid (14:0) and palmitic acid (16:0) were significantly lower in haemolymph of AKH-treated locusts than that in the FB. No significant effect on the relative percentage of the FA level between the hormones was recorded (t-test). The bug FB released mainly linoleic acid (18:2) after the Locmi-AKHs treatment (Figure 6B) and also the pattern of other FAs is similar to that shown in the Figure 4. As revealed by the conspecific assays, the FA with C>18 are not AKH mobilized. At least one of the Locmi-AKHs was responsible for a significantly higher level (t-test) of palmitoleic acid (16:1) and linolenic acid (18:3) in the treated haemolymph compared to their levels in the bug´s FB. The level of linoleic acid (18:2) was significantly lower after the treatment of all Locmi-AKHs than in the intact FB.

No significant effect on the relative percentage of FA levels between the hormones was recorded (one-way ANOVA, Tukey post test).

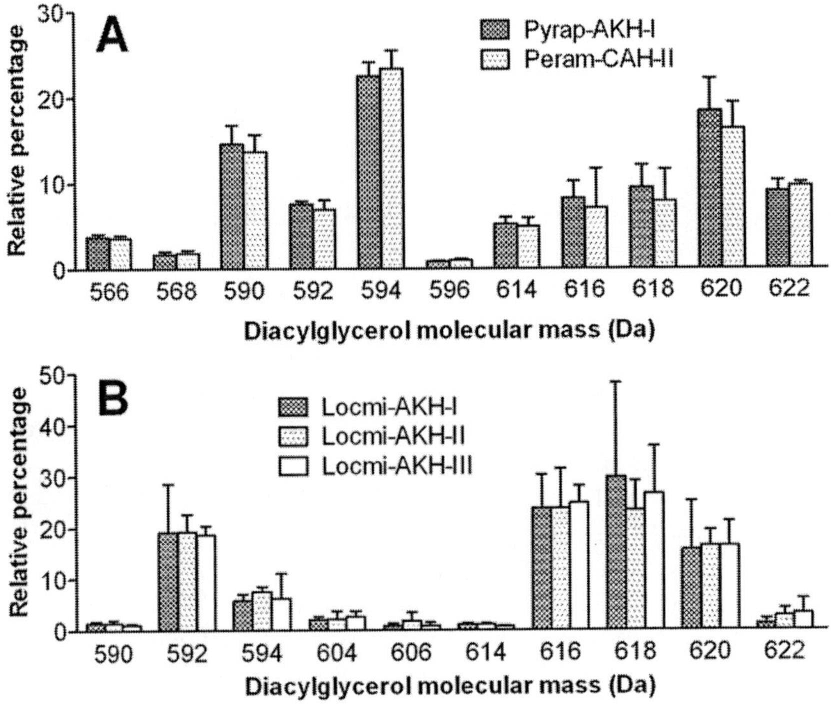

Figure 5. Heterologous assays. Relative percentage of the individual DG species participating in total elevation of haemolymph lipids (=100 %) of (A) *L. migratoria* after the treatment by bug AKHs Pyrap-AKH and Peram-CAH-II, and (B) *P. apterus* after the treatment by locust AKHs Locmi-AKH-I, AKH-II and AKH-II. Statistics: t-test (A: DG mobilized by Pyrap-AKH vs. DG mobilized by Peram-CAH-II) and one-way ANOVA (B: DG mobilized by Locmi-AKHs among themselves) revealed no significant differences at the 5% level. The bars represent SD, n=4.

The species-specific mobilization of the DGs/FAs is not so surprising, because there are a lot of examples in literature available showing the species-specific mobilization of different energy substrates elicited by AKHs. For example, while the Peram-CAH-II used in this study is known to mobilize carbohydrates in the cockroaches *P. americana* and *Blatta orientalis* (Witten et al., 1984; Scarborough et al., 1984), in the Colorado potato beetle *Leptinotarsa decemlineata* it mobilizes carbohydrates and induces proline synthesis (Gäde & Kellner, 1989), and in *P. apterus* it is involved in the mobilization of lipids only

(Kodrík et al., 2002). In the comparative heterologous study (Socha et al., 2004), it was proven that injections of Locmi-AKH-I and Tenmo-HrTh into the *P. apterus* body, and Pyrap-AKH, Peram-CAH-II and Tenmo-HrTh into the *L. migratoria* body mobilize lipids as is typical for the native AKHs of those species. On the other hand, injection of Locmi-AKH-I, Pyrap-AKH and Peram-CAH-II into the *Tenebrio molitor* body stimulates mobilization of carbohydrates as is typical for this beetle native Tenmo-HrTh. The question of whether the usage of different energy substrates in different insect species with the same structure of native AKH is coupled with evolution of different species-specific types of AKH receptors responsible for various signal transduction events remains to be answered.

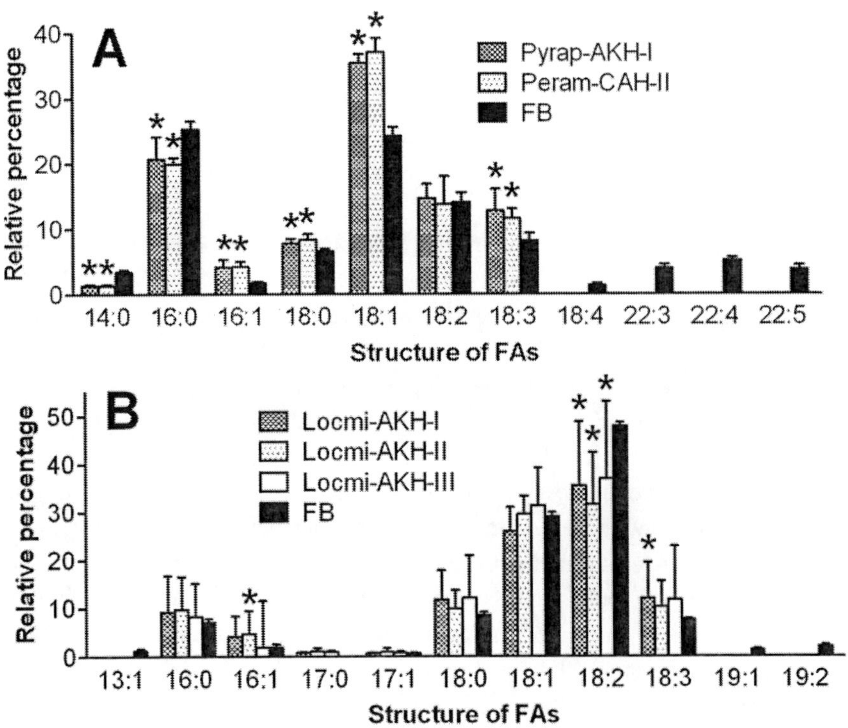

Figure 6. Heterologous assays. Relative percentage of the FAs contained in the haemolymph DGs mobilized by (A) Pyrap-AKH and Peram-CAH-II injected into the *L. migratoria*, and by (B) Locmi-AKHs injected into the *P. apterus*, and their comparison with FAs in the corresponding intact FB TGs. Statistically significant differences at the 5% level evaluated by t-tests (FAs in haemolymph DGs vs. FAs in FB TGs) are indicated by asterisks. The bars represent SD, n=4 for haemolymph samples, and n=10 (A; *L. migratoria*) and n=12 (B; *P. apterus*) for FB samples.

The results of this chapter showed that similar specificity seems to be obvious on the DG/FA level. On the other hand, this statement does not explain a different efficiency and selectivity of native AKHs on the mobilization of certain DGs and/or FAs. A biochemical mechanism of the specificity remains to be elucidated - one can just speculate that the AKH action indicates a lipase-specificity in insects, but the mechanism is unknown.

Conclusion

The HPLC/ESI-MS investigation of lipids mobilized from the FB into the haemolymph of *L. migratoria* and *P. apterus* after the injection with AKHs revealed the following matters:

1. The mobilization of DGs and/or FAs after the injection of native AKHs is not uniform. The individual DG species are mobilized selectively, preferably with the DGs containing the C18 and C16 FAs, especially the C18 unsaturated FAs.
2. There is a different spectrum of the FAs in the FB TGs and in the haemolymph DGs mobilized by injection of the AKHs, thus, the percentual pattern of the FAs in mobilized DGs does not directly reflect that in the FB TGs. This indicates that some FAs are preferred.
3. The FAs with a carbon chain longer than 18 and shorter than 14 do not participate in lipid mobilization, although they are present in the FB.
4. In *L. migratoria,* all Locmi-AKHs predominantly mobilize oleic acid (18:1), moreover unsaturated linoleic acid (18:2) is preferably mobilized by AKH-I, while saturated stearic acid (18:0) and partially myristic acids (14:0) by AKH-II and AKH-III.
5. In *P. apterus,* both innate hormones mobilize mainly linoleic acid (18:2). Nevertheless the mobilization is significantly more intensive under the thumb of Pyrap-AKH, while the Peram-CAH-II is significantly more active in mobilization of oleic acid (18:1).
6. Heterologous assay, when the *L. migratoria* AKHs were injected into the *P. apterus* body and vice versa, the *P. apterus* AKHs into the *L. migratoria* body reveled that the lipid mobilization pattern is rather species-specific and mostly depending on a character of the target FB.

In addition to those statements, the study has opened a few questions. First of all, it is the biochemical background of the AKH selectivity towards the mobilized DGs of FAs. Nothing is known about the phenomenon and an intensive research is necessary to elucidate it. Also, the reasons why the spectrum of preferred FAs between the species differs are not completely clear, although the answer will probably be in different species biology and life strategy, including food preference, living conditions or a way of locomotion. The study has also brought some arguments to reasoning why some insect species have more than one AKH, however, this phenomenon is too complex and complicated, and its explanation will probably include more physiological and maybe evolutional reasons.

Acknowledgments

This study was supported by the grants Nos. P501/10/1215 (DK) and P502/10/1734 from the Czech Science Foundation, and by project No. Z50070508 of the Institute of Entomology, funded by the Academy of Sciences of the Czech Republic. The authors thank Mrs. D. Rienesslová for her technical assistance.

References

Arrese, E. L. & Wells, M. A. (1994) Purification and properties of a phosphorylatable triacylglycerol lipase from the fat body of an insect, *Manduca sexta*. *Journal of Lipid Research*, 35, 1652-1659.

Arrese, E. L. & Wells, M. A. (1997) Adipokinetic hormone-induced lipolysis in the fat body of an insect, *Manduca sexta*: synthesis of sn-1,2-diacylglycerols. *Journal of Lipid Research*, 38, 68-76.

Arrese, E. L., Rojas-Rivas, B. I. & Wells, M. A. (1996) The use of decapitated insects to study lipid mobilization in adult *Manduca sexta*, effects of adipokinetic hormone and trehalose on fat body lipase activity. *Insect Biochemistry and Molecular Biology*, 26, 775-782.

Arrese, E. L., Canavoso, L. E., Jouni, Z. E., Pennington, J. E., Tsuchida, K. & Wells M. A. (2001) Lipoprotein metabolism in insects: current status and future directions. *Insect Biochemistry and Molecular Biology*, 31, 7-17.

Bártů, I., Tomčala, A., Socha, R., Šimek P. & Kodrík, D. (2010) Analysis of lipids mobilized by adipokinetic hormones in the firebug *Pyrrhocoris apterus* (L.) *European Journal of Entomology*, 107, 509-520.

Beenakkers, A. M. T. & Scheres, J. M. J. C. (1971) Dietary lipids and lipid composition of the fat body of *Locusta migratoria*. *Insect Biochemistry*, 1, 125-129.

Beenakkers, A. M. T., Van der Horst, D. J. & Van Marrewijk, W. J. A. (1985) Insect lipids and lipoproteins, and their role in physiological processes. *Progress in Lipid Research*, 24, 19-67.

Canavoso, L. E. & Wells, M. A. (2000) Metabolic pathways for diacylglycerol biosynthesis and release in the midgut of larval *Manduca sexta. Insect Biochemistry and Molecular Biology*, 30, 1173-1180.

Canavoso, L. E., Frede, S. & Rubiolo, E. R. (2004) Metabolic pathways for dietary lipids in the midgut of hematophagous *Panstrongylus megistus* (Hemiptera: Reduviidae). *Insect Biochemistry and Molecular Biology*, 34, 845-854.

Canavoso, L. E., Bertello, L. E., De Lederkremer, R. M. & Rubiolo, E. R. (1998) Effect of fasting on the composition of the fat body lipid of *Dipetalogaster maximus*, *Triatoma infestans* and *Panstrongylus megistus* (Hemiptera: Reduviidae). *Journal of Comparative Physiology*, 168 B, 549-554.

Chino, H. & Downer, R. G. H. (1982) Insect hemolymph lipophorin: a mechanism of lipid transport in insects. *Advances in Biophysics*, 15, 67-92.

Downer, R. G. H. (1985) Lipid metabolism. In G. A. Kerkut, & L. I. Gilbert (Eds.), *Comprehensive Insect Physiology, Biochemistry and Pharmacology, Vol. 10* (pp. 77-113). Oxford: Pergamon Press.

Ebberink, R. H. M., Smit, A. B. & Van Minnen, J. (1989) The insulin family, evolution of structure and function in vertebrates and invertebrates. *Biological Bulletin*, 177, 176-182.

Fernlund, P., Josefsson, L. (1972) Crustacean color-change hormone: Amino acid sequence and chemical synthesis. *Science*, 177, 173-175.

Gäde, G. (2009) Peptides of the adipokinetic hormone/red pigment-concentrating hormone family: a new take on biodiversity. *Annals of New York Academy of Sciences*, 1163, 125-136.

Gäde, G. & Auerswald, L. (2002) Beetles' choice - proline. *Comparative Biochemistry and Physiology B-Biochemistry & Molecular Biology*, 132, 117-129.

Gäde, G. & Auerswald, L. (2003) Mode of action of neuropeptides from the adipokinetic hormone family. *General and Comparative Endocrinology*, 132, 10-20.

Gäde, G. & Goldsworthy, G. J. (2003) Insect peptide hormones: a selective review of their physiology and potential application for pest control. *Pest Management Science*, 59, 1063- 1075.

Gäde, G. & Kellner, R. (1989) The metabolic neuropeptides of the corpus cardiacum from the potato beetle and the American cockroach are identical. *Peptides*, 10, 1287-1289.

Gäde, G., Hoffmann, K. H. & Spring, J. H. (1997) Hormonal regulation in insects: facts, gaps, and future directions. *Physiological Reviews,* 77, 963-1032.

Gäde, G. & Marco, H.G. (2006) Structure, function and mode of action of select arthropod neuropeptides. In A. U. Rahman (Ed.), *Studies in Natural Products Chemistry*, 33, 69-139.

Gäde, G., Šimek, P., Clark, K. D. & Auerswald, L. (2006) Unique translational modification of an invertebrate neuropeptide: a phosphorylated member of the adipokinetic hormone peptide family. *Biochemical Journal*, 393, 705-713.

Gäde, G., Auerswald, L., Šimek, P., Marco, H. G. & Kodrík, D. (2003) Red pigment concentrating hormone is not limited to crustaceans. *Biochemical and Biophysical Research Communication*, 309, 972-978.

Goldsworthy, G. J. (1990) Hormonal control of flight metabolism in locusts. In R. F. Chapman, & A. Joern (Eds.), *Biology of grasshoppers* (pp. 205-225). New York: John Wiley and Sons.

Goldsworthy, G. J. (1994) The insect adipokinetic hormones: are they the insect glucagons? In K. G. Davey, R. E Peter, & S. S. Tobe (Eds.), *Perspectives in Endocrinology: Proceedings of XII. International Congress of Comparative Endocrinology* (pp. 486-492). Ottawa: National Research Council of Canada.

Goldsworthy, G. J., Chandrakant, S. & Opoku-Ware, K. (2003) Adipokinetic hormone enhances nodule formation and phenoloxidase activation in adult locusts injected with bacterial lipopolysaccharide. *Journal of Insect Physiology*, 49, 795-803.

Goldsworthy, G. J., Opoku-Ware, K. & Mullen, L.M. (2002) Adipokinetic hormone enhances laminarin and bacterial lipopolysaccharide-induced activation of the prophenoloxidase cascade in the African migratory locust, *Locusta migratoria*. *Journal of Insect Physiology*, 48, 601-608.

Goldsworthy, G. J., Lee, M. J., Luswata, R., Drake, A. F. & Hyde, D. (1997) Structures, assays and receptors for locust adipokinetic hormones. *Comparative Biochemistry and Physiology*, 117B, 483-496.

Hansen, K. K., Hauser, F., Cazzamali, G., Williamson, M. & Grimmelikhuijzen C. J. P. (2006) Cloning and characterization of the adipokinetic hormone receptor from the cockroach *Periplaneta americana*. *Biochemical and Biophysical Research Communication*, 343, 638-643.

Hoffman, A. G. D. & Downer, R. G. H. (1979) End product specificity of triacylglycerol lipases from intestine, fat body, muscle and haemolymph of the American cockroach, *Periplaneta americana* L. *Lipids*, 14, 893–899.

Howard, R. W. & Stanley-Samuelson, D. W. (1996) Fatty acid composition of fat body and Malpighian tubules of the tenebrionid beetle, *Zophobas atratus*: significance in eicosanoid-mediated physiology. *Comparative Biochemistry and Physiology*, 115B, 429-437.

Kaufmann, C. & Brown, M. R. (2006) Adipokinetic hormone in the African malaria mosquito, *Anopheles gambiae*: Identification and expression of genes for two peptides and a putative receptor. *Insect Biochemistry and Molecular Biology*, 36, 466-481.

Kodrík, D. (2008) Adipokinetic hormone functions that are not associated with insect flight. *Physiological Entomology*, 33, 171-180.

Kodrík, D. & Goldsworthy, G. J. (1995) Inhibition of RNA synthesis by adipokinetic hormones and brain factor(s) in adult fat body of *Locusta migratoria. Journal of Insect Physiology*, 4, 127-133.

Kodrík, D., Krishnan, N. & Habuštová, O. (2007) Is the titer of adipokinetic peptides in *Leptinotarsa decemlineata* fed on genetically modified potatoes increased by oxidative stress? *Peptides*, 28, 974-980.

Kodrík, D., Šimek, P., Lepša, L. & Socha, R. (2002) Identification of the cockroach neuropeptide Pea-CAH-II as a second adipokinetic hormone in the firebug *Pyrrhocoris apterus*. *Peptides*, 23, 585-587.

Kodrík, D., Socha, R., Šimek, P., Zemek, R. & Goldsworthy, G. J. (2000) A new member of the AKH/RPCH family that stimulates locomotory activity in the firebug, *Pyrrhocoris apterus* (Heteroptera). *Insect Biochemistry and Molecular Biology*, 30, 489-498.

Kodrík, D., Marco, H. G., Šimek, P., Socha, R., Štys, P. & Gäde G. (2010) *The adipokinetic hormones of Heteroptera: A comparative study. Physiological Entomology*, 35, 117-127.

Köllisch, G. V., Lorenz, M. W., Kellner, R., Verhaert, P. D. & Hoffmann, K. H. (2000) Structure elucidation and biological activity of an unusual adipokinetic hormone from corpora cardiaca of the butterfly, *Vanessa cardui. European Journal of Biochemistry*, 267, 5502-5508.

Lorenz, M. W. (2003) Adipokinetic hormone inhibits the formation of energy stores and egg production in the cricket *Gryllus bimaculatus. Comparative Biochemistry and Physiology B*, 136, 197-206.

Martin, J. S. (1969a) Lipid composition of fat body and its contribution to the maturing oocytes in *Pyrrhocoris apterus*. *Journal of Insect Physiology*, 15, 1025-1045.

Martin, J. S. (1969b) Studies on assimilation, mobilization, and transport of lipids by the fat body and haemolymph of *Pyrrhocoris apterus*. *Journal of Insect Physiology*, 15, 2319-2344.

Murata, M. & Tojo, S. (2002) Utilization of lipid for flight and reproduction in *Spodoptera litura* (Lepidoptera: Noctuidae). *European Journal of Entomology,* 99, 221-224.

Oudejans, C. R. H. M., Dijkhujsen, R. M., Kooiman, F. P. & Beenakkers, A. M. T. (1992) Dose-response relationship of adipokinetic hormones (Lom-AKH-I, II and III) from the migratory locust, *Locusta migratoria. Proceedings of the Section: Experimental and Applied Entomology of the Netherlands Entomological Society*, 3, 165-166.

Oudejans, R. C. H. M., Kooiman, F. P., Heerma, W., Versluis, C., Slotboom, A. J. & Beenakkers A. M. T. (1991) Isolation and structure elucidation of a novel adipokinetic hormone (Lom-AKH-III) from the glandular lobes of the corpus cardiacum of the migratory locust, *Locusta migratoria. European Journal of Biochemistry*, 195, 351-359.

O'Shea, M., Witten, J. & Schaffer, M. (1984) Isolation and characterization of two myoactive neuropetides: further evidence for an invertebrate peptide family. *Journal of Neuroscience*, 4, 521- 529.

Park, Y., Kim, Y. J. & Adams, M .E. (2002) Identification of G protein-coupled receptors. *Proceedings of the National Academy of Science of the United States of America*, 99, 11423-11428.

Rangan, V. S. & Smith, S. (2002) Fatty acid synthesis in eukaryotes. In D. E. Vance, & J. E. Vance (Eds.), *Biochemistry of lipids, lipoproteins and membranes* (pp. 151-180). Paris: Elsevier Science B.V.

Satake, H., Takuwa, K., Minakata, H. & Matsushima, O. (1999) Evidence for conservation of the vasopressin/oxytocin superfamily in annelids. *Journal of Biological Chemistry*, 274, 5605-5611.

Scarborough, R. M., Jamieson, G. C., Kalisz, F., Kramer, S. J., McEnroe, G. A., Miller, C. A. & Schooled, D. A. (1984) Isolation and primary structure of two peptides with cardio acceleratory and hyperglycaemic activity from the corpora cardiaca of *Periplaneta americana. Proceedings of the National Academy of Science of the United States of America,* 81, 5575-5579.

Scharrer, B. & Scharrer, E. (1944) Neurosecretion. IV. Comparison between the intercerebralis-cardiacum-allatum system of the insects and the hypothalamo-hypophyseal system of the vertebrates. *Biological Bulletin,* 87, 242-251.

Schneider, M. & Dorn, A. (1994) Lipid storage and mobilization by flight in relation to phase and age of *Schistocerca gregari*a females. *Insect Biochemistry and Molecular Biology*, 24, 883-889.

Siegert, K. J., Morgan, P. J. & Mordue, W. (1985) Primary structure of locust adipokinetic hormone II. *Biological Chemistry Hoppe-Seyler*, 366, 723-727.

Socha, R. (1993) *Pyrrhocoris apterus* (Heteroptera) – an experimental model species: a review. *European Journal of Entomology,* 90, 241-286.

Socha, R. & Kodrík, D. (1999) Differences in adipokinetic response of *Pyrrhocoris apterus* (Heteroptera) in relation to wing dimorphism and diapause. *Physiological Entomology*, 24, 278-284.

Socha, R. & Šula, J. (2008) Regulation of the development of flight muscles in long-winged adults of the flightless bug, *Pyrrhocoris apterus* (Heteroptera: Pyrrhocoridae*). European Journal of Entomology*, 105, 575-583.

Socha, R. & Zemek, R. (2000) Wing movement behavior in long- and short-winged morphs of the flightless bug *Pyrrhocoris apterus* L. (Heteroptera: Pyrrhocoridae*). Journal of Insect Behaviour*, 13, 741-750.

Socha, R. & Zemek, R. (2003) Wing morph-related differences in the walking pattern and dispersal in a flightless bug, *Pyrrhocoris apterus* (L.) (Heteroptera). Oikos, 100, 35-43.

Socha, R., Kodrík, D. & Šula, J. (2005) Wing morph-specific differences in the metabolism and endocrine control of reserve mobilization in adult males of a flightless bug, *Pyrrhocoris apterus* (L.) (Heteroptera). *Journal of Comparative Physiology*, 175 B, 557-565.

Socha, R., Kodrík, D. & Zemek, R. (1999) Adipokinetic hormone stimulates insect locomotor activity. *Naturwissenschaften*, 88, 85-86.

Socha, R., Kodrík, D., Šimek, P. & Patočková, M. (2004) The kind of AKH-mobilized energy substrates in insects can be predicted without a knowledge of the hormone structure. *European Journal of Entomology,* 101, 29-35.

Soulages, J. L. & Wells, M. A. (1994) Lipophorin: The structure of an insect lipoprotein and its role in lipid transport in insects. *Advances in Protein Chemistry*, 45, 371-415.

Staubli, F., Jorgensen, T. J. D., Cazzamali, G., Williamson, M., Lenz, C., Sondergaard, L., Roepstorff, P. & Grimmelikhuijzem, C. J. P. (2002) Molecular identification of the insect adipokinetic hormone receptors. *Proceedings of the National Academy of Science of the United States of America*, 99, 3446-3451.

Stone, J. V., Mordue, W., Blatney, K. E. & Morris, H. R. (1976) Structure of locust adipokinetic hormone, a neurohormone that regulates lipid utilization during flight. *Nature*, 263, 207-211.

Tomčala, A., Bártů, I., Šimek, P. & Kodrík D. (2010) Locust adipokinetic hormone mobilizes diacylglycerols selectively. *Comparative Physiology and Biochemistry* 156 B, 26-32.

Turunen, S. & Crailsheim, K. (1996) Lipid and sugar absorption. In M. J. Lehane, & P. F. Billingsley (Eds.), *Biology of Insect Midgut* (pp. 293-320). London: Chapman and Hall.

Van der Horst, D. J. (1982) Lipid transport in insects. In T. E. Mittler, & R. H. Dadd (Eds.), *Metabolic Aspects of Lipid Nutrition in Insects* (pp. 183-202). Colorado: Westview Press, Boulder.

Van der Horst, D. J., Van Marrewijk, W. J. A. & Diederen, H. B. (2001) Adipokinetic hormones of insect: release, signal transduction, and responses. *International Review of Cytology*, 211, 179-240.

Witten, J. L., Schaffer, M. H., O'Shea, M., Cook, J. C., Hemling, M. E. & Rinehart, Jr., K. L. (1984) Structures of two cockroach neuropeptides assigned by fast atom bombardment mass spectrometry. *Biochemical and Biophysical Research Communication*, 124, 350-358.

In: Saturated Fats
Editor: Joseph P. Langella
ISBN: 978-1-61209-566-0

Chapter V

Lipid Composition of Erythrocytes at Cardiovascular and Hepatobiliary Diseases

Natalia V. Zhukova[1] and Tatyana P. Novgorodtseva[2]
[1] Institute of Marine Biology, Far East Branch of Russian Academy of Sciences, Vladivostok, Russia
[2] Institute of Medical Climatology and Rehabilitative Treatment, Siberian Branch of Russian Academy of Medical Sciences, Vladivostok, Russia

Introduction

Composition of human tissues lipids and their metabolism are of paramount importance for the growth, development and long-term human health. Lipids are the major source of energy in the body and supply essential lipid-soluble vitamins and polyunsaturated fatty acids (PUFA) that are required in relatively high amounts during growth and life. Lipids affect the composition of membrane structures and modulate membrane functions as well as the functional development of the central nervous system. Some long-chain PUFA serve as precursors for bioactive lipid mediators.

The principal long-chain polyunsaturated fatty acids, arachidonic acid (20:4n-6, AA), eicosapentaenoic acid (20:5n-3, EPA), and docosahexaenoic acid (C22:6n-3, DHA), i.e. products of chain elongation and desaturation of the essential fatty acids, linoleic acid (18:2n-6, LA) and α-linolenic acid (18:3n-3, ALA), are important for growth and maintaining various biological functions in the human organism (Nagao, Yanagita, 2005). Long chain PUFA of the n-3 series are essential for early development of the central nervous system (Uauy et al., 2003; Desci, Koletzko, 2000), whereas at later stages of life, they are beneficial for cardiovascular health, reducing platelet aggregation and triacylglycerol (TAG) levels; they may play a role in the prevention of coronary artery disease (Caggiula, Mustad, 1997), hypertension and arthritis, as well as some inflammatory and autoimmune disorders (Calder, 2001) and some cancers (Bagga et al., 1997; Pala et al; 2001). EPA, dihomo-γ-linolenic acid (20:3n-6) AA, and DHA are precursors of prostaglandins, thromboxanes and leukotrienes, which are powerful regulators of numerous cell functions, playing positive role in the pathophysiology of thrombosis and atherosclerosis, as well as immunologic processes and inflammation. The percentage of EPA + DHA in red blood cells, also known as the Omega-3 Index, has recently been proposed as a risk marker for death from coronary heart disease.

A prospective direction in the research of fatty acids is connected with the influence on their composition and metabolism of hereditary traits and environmental factors, as well as with the observation that the composition of blood fatty acids has information value for a broad range of diseased states.

Changes in the fatty acid composition associated with distrurbances of metabolism are characteristic of many therapeutic noninfectious diseases. Deviation from the ormal values (when compared to healthy subjects) is of an important diagnostic value. A relationship between disorders of lipid exchange and pathophysiology of cardiovascular and other therapeutic diseases have been known for a long, but still many aspects of this problem are investigated insufficiently.

A large body of research data proves that the fatty acid composition of different blood fractions may be used for evaluation of the effect of lifestyle variables and diseases on metabolic processes in organism. The abnormalities in the phospholipid and fatty acid composition in erythrocyte membranes and plasma are recognized as a key to the diagnosis of the some therapeutic diseases. The fatty acid composition of the erythrocyte membrane is the integrated result of interactions between dietary fatty acid intake and endocrine changes (Brenner, 1989), partly because of the long half-life (120 days) of erythrocytes. The parameter may, therefore, be an appropriate biomarker for investigating not only dietary intake but also it may better reflect the fatty acid status of the organism than the fatty acid composition of plasma lipids.

A lipid analysis of erythrocyte membranes is used as a diagnostic tool for various diseases including chronic fatigue syndrome (Nijs, De Meirleir, 2004), Zillweger syndrome spectrum (Krause et al., 2009), schizophrenia (Fenton et al., 2000), clinical symptoms of depression (Adams et al., 1996). For example, it is hypothesized that oxidative stress accounts for the observed reduction in the levels of essential fatty acids in the erythrocyte membrane of chronic fatigue syndrome patients (Nijs, De Meirleir, 2004). Erythrocyte fatty acid composition is associated with the metabolic syndrome and its components (Kabagambe et al., 2008). The relationships have been reported between erythrocyte membrane fatty acids and postmenopausal breast cancer risk (Pala et al., 2001).

As a result, alterations of the fatty acid profile of erythrocyte membranes result in changes in the degree of unsaturation of lipids, in the antioxidant levels of the erythrocyte membrane, in decrease of both membrane fluidity, and the activity of membrane-bound enzymes. Because the erythrocyte membrane serves as a variable barrier to oxygen transport, changes in its composition can induce cellular hypoxia in the tissue. Furthermore, because the size, shape, and diffusion capacity of a red blood cell depend on the structure of its membrane, alterations in the membrane structure could lead to a decrease in tissue oxygenation (Buchwarld et al., 2000). Such modifications of the red cell membrane in cardiovascular cells might be a part of a pathogenetic mechanism responsible for pathologic states, such as atherosclerosis and hypertension.

The liver plays a central role in the metabolism of PUFA. Liver function influences the homeostasis of complex lipids in plasma (Arranz et al., 1996). Liver disorders are accompanied by major quantitative and qualitative changes in plasma lipoprotein metabolism (Taus et al., 1999). Thereby, investigation of the therapeutic diagnostic potential of the lipids in hepatobiliary diseases is especially important. Alterations in plasma lipoprotein composition and erythrocyte membranes have been observed in different liver pathologies (e.g. Robberecht et al., 1997; Socha et al., 1997; Malinska et al., 2009; Sumida et al., 2009).

Many studies revealed changes in the fatty acid content and lipid metabolism disturbances, but data on effects of the diseases on the PUFA status and lipid metabolism are contradictory.

In this chapter, we explore the role of the changes in the fatty acid composition of erythrocyte membranes and metabolic disturbances at the development of cardiovascular and hepatobiliary diseases. Special attention has been given to the investigation of the characters of serum lipids and erythrocyte membrane lipids that may be used for diagnosis of therapeutic diseases.

Serum biochemical characterisitics and data on erythrocyte membrane lipids serve as potential indicator for assessment of the health status and disease severity and may be used as an important diagnostic tool in the medical practice.

The Diagnostic Information Value of Human Blood Fatty Acids

Fatty acids play an important part in the development of noninfectious diseases, metabolism is affected by hereditary traits and environmental factors, and that the composition of blood fatty acids has information value for a broad range of diseased states. A large body or research data proves that the fatty acid composition of different blood fractions may be used for evaluation of the effect of lifestyle variables on metabolic processes. The diagnostic significance of blood fatty acid composition is reported by many authors (Harwood, 1994; Warensjo et al., 2006; Jakobik et al., 2009).

Alteration of the blood fatty acid composition may be caused by lipid metabolism disturbances, such as a decrease in desaturase activity impairing the biosynthesis of long chain fatty acids; an increase in the degradation, particularly peroxidation, of polyunsaturated fatty acids (PUFA); and an enhancement of β-oxidation of the fatty acids (Harwood, 1994). In the past decade, the assumption (hypothesis) becomes widespread about endogenous assemblage of an active transport of the PUFA into the cells, which blockage contributes to the development of disease, (Titov, 1998). The endogenous PUFA intake into the cells is particularly important because the lack of these essential fatty acidss causes the development of diseases. The chronic endogenous deficiency of n-3 fatty acids in cells results in the alteration of the fatty acid composition of phospholipids and changes physical and chemical properties of the plasma cell membranes. The decrease in cell membrane fluidity inhibits the activity of Na, K ATP-ase and of insulin receptors. The n-3 fatty acid deficiency in the cells is one of the major pathogenetic factors in developing hyperlipoproteinemia, hypertensia, thrombosis, and diabetes mellitus.

It has been found that one of the pathogenetic mechanisms of diseases associated with atherosclerosis is the changes in the structure and functions of erythrocyte membranes. The changes in the phospholipid composition of the biological membranes and in the distribution of the lipids between the membrane leaflets can be a cause of physiological modifications (Yamaji-Hasegama, Tsujimoto, 2006). A majority of fatty acids in red blood cells are esterified in phospholipids, and the distribution of fatty acids between individual phospholipids is distinctly phospholipid specific and significantly different from their distribution in plasma. In erythrocytes phosphatidylethanolamine (PE) and phosphatidylserine (PS) are more unsaturated in comparison with phosphatidylcholine (PC), however, PS contains up to 40% of stearic acid (18:0) and from 10 to 25% of palmitic acid (16:0). The fatty acid composition of membrane lipids determines physical and chemical properties of the membranes. The increase of the hydrocarbon chain lengths or decrease of the degree of the unsaturation of the fatty acids in erythrocyte phospholipids leads to the decrease in membrane fluidity (Salem, Niebylski, 1995). Mature red blood cells are unable to produce fatty acids. The composition of erythrocyte fatty acids results either from uptake of dietary fatty acids from plasma lipids or from their synthesis de novo in erythrocytes (Harwood, 1994; Poppitt et al., 2005). Biological membranes are composed of phospholipids with characteristic fatty acid compositions, which ensure the functional specificity of the membranes.

It is to be noted that most authors studying the significance of fatty acid modification and disturbances in their metabolism on the development of diseases take into account the fatty acids in different lipid fractions of blood serum and plasma. The use of the fatty acids of blood cells mainly of erythrocytes, is uncommon in the diagnostic investigations. Nevertheless, the fatty acid composition of erythrocyte membrane lipids may better reflect the fatty acid status of the organism than the fatty acid composition of plasma lipids. This consideration defines the information value of erythrocyte fatty acids. Change in the fatty acid composition of erythrocyte membranes is a more sensitive indicator of the lipid metabolism disturbance, and erythrocytes can serve as a unique model for analyzing fatty acids. According to many reports, the analysis of long chain fatty acids in red blood cells gives more reliable and stable results than analysis of highly varying fatty acid compositions of serum and plasma lipids (Novgorodtseva, Endakova, 1993; Lausada et al., 2007; Jakobik et al., 2009). A comparative analysis of the lipid composition of the atherosclerotic plaque, plasma and erythrocyte membrane in patients with advanced coronary heart disease allowed the authors to suggest that analysis of the lipid composition of erythrocyte membrane could provide a useful tool to monitor the evolution of the coronary heart disease (Lausanda et al., 2007). The assessment of n-3 long-chain PUFA status in humans showed that plasma, plasma phospholipid, plasma TAG, plasma cholesteryl ester, erythrocyte, erythrocyte phospholipid, and platelet were all effective biomarkers of DHA status in adult men and women (Fekete et al., 2009).

Fatty acid composition patterns of erythrocyte membrane predict development of Type 2 diabetes mellitus (Krachler et al., 2008). Higher proportions of pentadecanoic acid (15: 0) and heptadecanoic acid (17:0) are associated with a lower risk of diabetes. In accordance with earlier findings, higher proportions of palmitoleic (16:1n-7), dihomo-γ-linolenic (20:3n-6) and adrenic (22:4 n-6) acids are associated with increased risk, whereas linoleic (18:2n-6) and clupanodonic (22:5n-3) acids are inversely associated with diabetes risk. The only significant predictors are 15:0 and 17:0 as protective factors and 22:4n-6 as risk factor.

Lipid composition of erythrocyte membranes was used as diagnostic criteria for assessment of different diseases including the chronic fatigue syndrome (Nijs, De Meirleir, 2004), Zillweger syndrome spectrum (Krause et al., 2009), schizophrenia (Fenton et al., 2000), clinical symptoms of depression (Adams et al., 1996). Fatty acid composition of erythrocyte membranes change at therapeutic diseases, such as chronic bronchitis (Novgorodtseva et al., 2006) and comorbidities of viscera (Shestopalov et al., 2009). It is hypothesized that oxidative stress accounts for the observed reduction in the levels of essential fatty acids in the erythrocyte membrane of chronic fatigue syndrome patients (Nijs, De Meirleir, 2004). Erythrocyte fatty acid composition is associated with the metabolic syndrome and its components (Novgorodtseva et al., 2008; 2009; Kabagambe et al., 2008). Total PUFA content, n-6 fatty acids specifically, are inversely associated, whereas saturated fatty acids are positively associated with the metabolic syndrome. The relationships between erythrocyte membrane fatty acids and postmenopausal breast cancer risk have been shown (Pala et al., 2001). Analysis suggests an inverse association between total PUFA and breast cancer risk, but individual PUFA behave differently. There is no association between saturated fatty acids and breast cancer risk. Monounsaturated fatty acids, mainly oleic acid, and saturation index in erythrocyte membranes are predictors of postmenopausal breast cancer risk.

Relation between the fatty acid composition of erythrocyte membranes and the lipid composition of blood serum, which is (being) a marker of atherosclerosis risk, has been confirmed. Theret et al. (1993) established the relationship between the phospholipid fatty acid composition of red blood cells, plasma lipids and apolipoprotiens (apo) B for 45–66 year old men. The level of docosahexaenoic acid (DHA) 22:6n-6 in erythrocyte phospholipids positively correlated with the contents of total cholesterol (ST), low density lipoprotein (LDL) cholesterol, and apo B in plasma. There is a definite positive dependence between α-linolenic acid (ALA) 18:n-3, apo A-1 and HDL cholesterol, and negative dependence between arachidonic acid (AA) 20:4n-6 in erythrocytes and plasma ST content.

Concentrations of palmitic and stearic acids are inversely related to apo B content. The literature on the subject demonstrates that analysis of changes in fatty acid composition in erythrocyte membranes is an important tool for detecting ontogenetic variations of lipid content in nerve and brain tissues and in retina (Salem et al., 1995). It has been generally recognized that DHA level in erythrocytes is a marker of the n-3 fatty acid status, and linoleic acid (LA) 18:2n-6 level is a marker of the n-6 fatty acid status. It is DHA in erythrocytes that is known to reflect accumulation of neural DHA with nerve tissue development (Innis, Owen, 2001). Arachidonic acid (AA) and DHA in cell membranes stimulate psychomotor development of children, as well as the development of their visual function. DHA content in red blood cell phospholipids positively correlates with psychomotor development indices in premature infants (van Wezel-Meijler et al., 2008). Reviewing the literature confirms high information value of the fatty acid composition of erythrocytes and the role of the lipid disturbances in the development of noninfectious pathologies, as well as substantiates the use of erythrocyte membranes for the assessment of the effect of exogenous and endogenous factors on the development of diseases.

It is known that the fatty acid composition of human blood reflects to a certain extent not only the fatty acid composition of dietary fat, but also the endogenous fatty acid synthesis, where desaturases play an important part. A surrogate measure of Δ9-, Δ6- and Δ5- desaturase activity can be calculated as the ratio of fatty acid product to precursor. A similar approach has been used for estimation of desaturase activity serving as markers of obesity and lifestyle variables and gender differences (Warensjo et al., 2006).

However, the authors often do not specify age and gender composition of the subject and control groups. Taking into consideration that the disturbance of fatty acid metabolism results in the alterations in the fatty acid composition, we find it justified to use the characteristic features of the fatty acid composition of erythrocytes as diagnostic criteria. The ratios of n-6 to n-3 PUFAs and the activity of enzymes, elongases and desaturases, participating in the biosynthesis of the fatty acids may serve as diagnostic criteria. In this approach, 20:4n-6/22:6n-3 and (20:3n-6+20:5n-3)/22:6n-3 ratios show the dependence between the fatty acid precursors and the inhibitors of eicosanoid synthesis.

In the present work, the 20:5n-3/22:5n-3 and 20:4n-6/22:4n-6 ratios were used to evaluate elongase activity. The ratio of 18:4n-3/18:3n-3 served as an indicator of Δ6 desaturase activity, whereas 20:4n-6/20:3n-6 ratio was used as an indicator of Δ5 desaturase activity. The concept on DHA biosynthesis suggests that the last step of DHA biosynthesis is characterized by the 22:6n-3/22:5n-3 ratio. The fatty acid metabolism disturbances can generally be evaluated by unsaturation index (UI) calculated as the sums of product of double bond in each fatty acid multiplied by the percentage of the total fatty acids (Galland, 1986; Ricci et al., 1987).

Fatty acids play a great part in the organism; this is confirmed by their presence in all live cells, by the high intensity of their metabolism, and their participation as precursors in the production of physiologically active derivatives, oxilipins. Fatty acids of phospholipids serve are essential structural constituents of the biological membranes. The fatty acid composition of the cells is affected by genetic and environmental factors. It is explains the high information value of fatty acid composition of erythrocyte membranes for biochemical and medical research.

Fatty Acid Composition of Blood Lipids: Age-Gender Aspects

Specific Local Dietary or Environmental Factors

The lipid metabolism is shown to distinguish significantly populations of different climatic conditions and geographical environments (Hegele, 1996). Some peculiarities in lipid metabolism and lipid peroxidation were shown for the population of Primorye region (The South of the Russian Far East. Lipid level in the serum of the 1153 healthy men from the population of the South Far East was shown to be similar with that of healthy men from Russia and Commonwealth of Independent States (CIS) (Table 1). Nevertheless, this characteristic differs only slightly from that of the native inhabitants of coastal and continental regions. Insignificant variations in the rate of lipid metabolism between populations of different regions of Russia may be caused by specific local dietary, genetic, or environmental factors.

Attempts to obtain normal reference values of the fatty acid composition of erythrocyte membrane lipids for investigations of disease-related alterations date back at least to the early 1970s (Cooper, 1970).

To determine the reference values and regional features of the fatty acid composition of erythrocyte membranes, 1153 healthy Vladivostok city residents were examined, they belonged to different age and gender groups living in this climatic zone for more than 10 years. An analysis of fatty acid composition was performed by gas-liquid chromatography (GLC) of methyl esters of fatty acids (MEFA) prepared from erythrocyte membranes of the individuals examined. A total of 48 fatty acids with carbon chain lengths from C12 to C22, even and odd, normal and branched, monounsaturated and polyunsaturated fatty acids (PUFA), were identified in the samples. The major fatty acids found in the erythrocytes of healthy subjects were palmitic 16:0, stearic 18:0, oleic 18:1n-9, linoleic 18:2n-6, arachidonic 20:4n-6 and docosahexaenoic 22:6n-3. The fatty acid compositions of erythrocytes in healthy adults are shown in Table 2 based on our results and literature data.

Table 1. Blood lipids of health men from different climatic and geographic regions, mean ± SE

Parameter, mmole·L^{-1}	Tallinn city*	Frunze city*	Russia*	Russian Far East (own data) n=1153
Total cholesterol	5.0 ± 0.1	5.4 ± 3.6	≤ 5.2	4.79 ± 0.51
Triacylglycerols	1.04 ± 0.05	1.68 ± 7.3	0.50–2.20	1.00 ± 0.11
HDL cholesterol	1.20 ± 0.03	1.70 ± 1.10	1.00–2.00	1.16 ± 0.13
LDL cholesterol	3.31 ± 0.08	3.4 ± 3.5	2.50–3.36	2.43 ± 0.59
VLDL cholesterol	0.49 ± 0.02	0.62 ± 1.50	0.27–0.36	0.25 ± 0.03

* cite by Endakova et al., 2002

Population and individual differences in the composition and metabolism of fatty acids were revealed for the inhabitants of different geographical regions. The significant differences in the fatty acid composition of erythrocyte membrane phospholipid and plasma triglycerides have been found between Indian Asians compared with Europeans were found (Peterson et al., 1994). Erythrocytes in Indian Asians contain high concentrations of linoleic, dihomo-γ-linolenic, and arachidonic acids, and low concentrations of oleic and n-3 fatty acids; plasma triglycerides contained higher linoleic and lower oleic acid levels. The authors denote a higher frequency of coronary heart disease in Indian Asians compared to Europeans and suggest that a high dietary intake of linoleic acid may not be cardioprotective (Peterson et al., 1994). High rates of coronary disease appear to be common to South Asia groups of different geographic origin, religion, and language. Low plasma high-density lipoprotein cholesterol, high plasma triacylglycerols, and high prevalence of type 2 diabetes have been consistently found in Asian Indians (Rajeshwari et al., 2005; Tziomalos et al., 2008). The study carried out in 11 European countries shown the differences between regions, with the highest percentages of saturated palmitic acid measured in Finnish students, the highest percentage of linolic acid in Belgium and the high contents of arachidonic acid in Southern Europe. Palmitic, oleic, linoleic and arachidonic acids measured in plasma have been related to plasma lipids and lipoproteins and to life-style variables including the body-mass index, alcohol consumption, tobacco and physical activity (Rosseneu et al., 1994).

Table 2. Fatty acid composition (%) of erythrocyte membranes of the healthy subjects, mean ± *SE or range.*

Fatty acid	Russian Far East (own data) n=1153	Manku et at, 1983*	Bougie et al, 1986*	Korf et al., 1987*	Samsonov et al., 1994*	Pita et al., 1997*
14:0	0.68 ± 0.03		1.1 ± 0.2	2.08	2.08	
15:0	0.41 ± 0.02				2.18	
i 6:0	1.26 ± 0.17					
16:0	19.26 ± 0.26	20.68 ± 0.26	22.2 ± 0.8	20.34	20.34	23.53 (22.26-25.75)
16:1	1.28 ± 0.09		1.7 ± 0.3	3.65	3.65	0.24 (0.06-0.39)
17:0	0.62 ± 0.05			2.40	2.40	
i 18:0	1.84 ± 0.23					
18:0	15.16 ± 0.26	14.71 ± 0.22	12.9 ± 0.2	11.95	11.95	17.20 (16.20-21.61)
18:1n-9	14.19 ± 0.25	9.78 ±0.24	21.2 ± 0.8	13.80	13.80	16.27 (14.95-18.62)
18:2n-6	11.01 ± 0.26		14.1 ± 1.0	11.77	11.77	9.54 (6.91-11.09)
18:3n-3	0.11 ± 0.02		0.7 ± 0.2	0.84	0.84	
18:4n-3	0.55 ± 0.11					
20:2n-6	0.67 ± 0.08					0.49 (0.38-0.74)
20:3n-6	1.05 ± 0.05	1.37 ± 0.05	2.1 ± 0.2	1.26	1.26	2.09 (1.71-2.68)
20:4n-6	10.81 ± 0.45	15.13 ± 0.29	18.4 ± 1.0	14.14	14.14	12.85 (11.92-14.78)
20:5n-3	1.55 ± 0.14	0.65 ± 0.03		1.20	1.20	0.67 (0.59-1.19)
22:4n-6	1.98 ± 0.16	5.54 ± 0.20				3.34 (2.39-4.22)
22:4n-3	0.16 ± 0.03					
22:5n-6	0.28 ± 0.08	3.99 ± 0.15				
22:5n-3	2.45 ± 0.07	2.53 ± 0.13		3.00	3.00	1.47 (1.36-1.89)
22:6n-3	8.32 ± 0.40	4.20 ± 0.15		4.47	4.47	5.06 (3.89-5.98)

*cited by Endakova et al., 2002

Differences in the regulatory mechanisms of fatty acid desaturation are the probable reason for the regional specific features of PUFA metabolism. The increased ability to produce PUFAs is found in Greenlan Eskimos and Caucasian Danes, as well as in American Indians from island Vancouver (Gibson, Sinclair, 1981; Horrobin, 1987). However metabolism of the fatty acids of these groups differs significantly by a low activity of Δ5 desaturase in Eskimos and Danes and low activity of Δ6 desaturase in American Indians. The elevated concentrations of

C20 and C22 fatty acids in blood lipids of the population of Primorye region may indicate an active enzyme of the last steps of the n-3 PUFA biosynthesis (Table 2.). These differences in lipid metabolism influence the epidemiologic situation in the population. Eskimos have a low rate of coronary disease in comparison with American Indians and the population of Russian Far East, which have the high level of the CVD. According to the data of Novgorodtseva and Endakova (1993) for the Russian Far East phenotype, the value of 22:6n-3/22:5n-3 ratio varies in healthy men in the range from 3.9 to 4.3. Growth of this ratio (underlies) indicate the risk of the CVD development.

Evaluation of the variability of the fatty acid composition of venous cord blood phospholipids in different populations of 11 European and 2 American countries shows that the amounts of saturated, monounsaturated and parent essential PUFA exhibit a relatively moderate variability among the data sets reviewed. Values of arachidonic acid and docosahexaenoic acid show a two-fold variability among the data sets. The highest values for docosahexaenoic acid are observed in areas with an apparently high consumption of dietary fat from sea fish (Minda et al., 2002).

Age- and Gender-Related Differences

Human gender and age, in addition to the environment and lifestyle, are important contributors to the formation of phenotype. Age- and gender-related differences in the lipid composition of blood fractions and of the other human organs and tissues are widely discussed in literature. Sex differences in blood cholesterol and fatty acid levels are frequently studied among groups of different ages, including prenatal development. It is shown that the levels of total cholesterol and high density lipoprotein (HDL) cholesterol are significantly higher in the female fetus than in the male one (Andoh et al., 1997). The same sex dependences in fetal organisms were detected for fatty acids, particularly for monoenoic fatty acid concentrations (Andoh et al., 1997). Circulating blood lipid concentrations also show gender-related differences (Williams, 2004). These differences are most marked in premenopausal women, in whom total cholesterol, LDL cholesterol and TAG concentrations are lower and HDL-cholesterol concentration is higher than in men.

Gender-related differences in the lipid composition of the organs and tissues of children were studied by several authors, but with inconsistent results. Sex differences in LDL cholesterol and HDL cholesterol levels were shown in children of school age (Lebedeva, 1994). The level of n-6 and n-3 PUFAs in erythrocyte lipids of 7–14-year-old boys differs from that of girls of the same age; boys have lower a C20 n-3 PUFA level. In contrast, Gerique with coauthors (1996) did not find out any sex differences in the levels of serum and plasma total cholesterol, total triglycerides, lipoprotein, HDL cholesterol and LDL cholesterol of Spanish children aged from 4 to 18 years. They revealed that plasma lipoprotein levels, associated with a high risk of cardiovascular disease (CVD) in childhood, remain unaltered until adulthood.

Analysis of the fatty acid composition of plasma phospholipids, triglycerides and cholesterol esters in healthy subjects from birth to young adulthood shows that after birth, percentage contributions of linoleic and α-linolenic acids increase, whereas long-chain PUFA decrease in all fatty acid fractions (Decsi, Koletzko, 1994). No difference in plasma fatty acid composition is detected between groups of children aged from 1 to 4, from 5 to 9, and from 10 to 15 years, except for a decrease in total saturated fatty acids of cholesterol esters between one and five and ten and fifteen years. No difference has been found in these values between boys and girls.

The visceral adipose tissue volume is positively correlated with TAG concentrations and negatively correlated with insulin-induced suppression of free fatty acids and insulin-mediated disposal of glucose (Montagne, O'Rahilly, 2000); therefore, this tissue is recognized as a marker of atherogenesis. The visceral adipose tissue volume is greater in men than in women (Sumner et al., 2002). Sex differences in the free fatty acid content of blood have been detected between African-American and Asian Indian men and women with insulin-dependent diabetes mellitus (Sumner et al., 1997). Compared with their female counterparts, African-American and Asian Indian men have higher TAG concentrations, weaker suppression of free fatty acid metabolism and higher risk of CVD (Sumner et al., 1997). The suppression of serum fatty acids decreases with an increase of the body mass index in men, whereas in women the level of fatty acids and TAG in serum does not depend on the body mass index. Nagaya et al. (1994) have proved that the interdependence between the fatty acid composition and cholesterol esters of serum differ in men and women. Stearic acid is negatively correlated with total cholesterol and total cholesterol /HDL cholesterol ratio in both men and women. In men myristinic acid (14:0) is negatively correlated with HDL cholesterol, and in women stearic acid (18:0) is positively correlated with HDL cholesterol. Palmitic acid (16:0) has no significant relation to any cholesterol indix in either sex (Nagaya et al., 1994). Many differences between men and women in metabolism of fatty acid and very low-density lipoprotein TAG and apo B-100 are reviewed in the study by Magkos et al. (2009).

An elevated level of such minor component as hexacosanoate 26:0 in erythrocyte membranes is significantly more frequent in male subjects than in females. The levels of erythrocyte membrane 26:0 are highly correlated with age. The frequency of high levels of erythrocyte membrane 26:0 in male subjects is greater than that in female subjects in all the seven age groups (Antoku et al., 2000).

Differences in lipid metabolism between men and women have been reported for decades (Jensen et al., 1996; Horton et al., 2002). In particular, the lipemic response to a high-fat meal is greater in men than in women. The elevated postprandial lipemia found in men compared with women is due to a prolonged availability of lipid chylomicrons, but is not related to differences in oxidation rates or incorporation of the ingested lipid into very low density lipoproteins (VLDL) in the liver (Knuth, Horowitz, 2006).

Ageing influences the fatty acid composition of adipose tissue independent of diet (Bolton-Smith et al. 1997). A comparison of the fatty acid composition of human adipose tissue in men and women aged 40-59 years reveals the relationship between individual fatty acids in adipose tissue and age, and between age and the ratio of linoleic acid to γ-linolenic as an indicator of Δ6 desaturase activity. Sex-consistent changes with age occurs for 18:2n-6 and 18:3n-6. These changes give rise to a significant increase in the 18:2n-6/18:3n-6 ratio with age. In addition, 20:3n-6 and sum of 22:5n-3 and 22:6n-3 also increases significantly with age. For the latter, the adjusted regression slopes are far greater for women than men. The detected decline in the rate limiting step of Δ6 desaturation occurs with age and is greater in women than in men. The sex differences may partially be due to inadequate adjustment for changes in the sex hormone status in males with ageing; this abnormality is probably regulated through diet.

Similarly to sex, age is an independent risk factor for CVD, atherosclerosis, and ischemic heart disease. Lipid metabolism disturbance is the most important determinant for atherogenesis. A majority of studies concerning age- and gender-related differences in the lipid composition of blood are focused on the changes in lipid metabolism.

Age-related changes in lipid and lipoprotein levels are revealed in healthy children and adults (Decsi, Koletzko, 1994; Boulton et al., 1995; Gerique et al., 1996; Lambert et al., 2004; Guerra et al., 2007), as well as in the subjects with burdened familial history, familial combined hyperlipidemia, and obesity (Shamir et al., 1996). In each age group, the relationships between lipid and lipoprotein profiles of human blood and food energy values, calcium contents in food, and fat consumption were traced (Boulton et al., 1995).

Changes in the fatty acid pool are observed in the ontogenetic development of the children. The fatty acid composition of phospholipids, triglycerides and cholesterol esters in plasma were determined in healthy subjects from birth to young adulthood (Decsi, Koletzko 1994). The contribution of linoleic and α-linolenic acids increases for several months after birth, whereas long-chain PUFA content decrease in all lipid fractions: phospholipids, triglycerides and cholesterol esters. No difference in the plasma fatty acid composition are seen between the groups of children aged 1-5, 5-10 and 10-15 years, except for a decrease of total saturated fatty acids in cholesterol esters. Relationships are shown to exist between maternal and neonatal essential fatty acid statuses. There is a relationship between maternal and neonatal essential fatty acid statuses and a significant correlation between the maternal and neonatal plasma n-6 to n-3 PUFA ratios (Al et al., 1996).

An investigation of the three-year tracking of fatty acid composition of plasma phospholipids in healthy children at ages 2, 3 and 5 years showed lower values for plasma monounsaturated fatty acids and the monounsaturated/saturated fatty acid ratio at 5 years compared to 2 years. In contrast, total PUFA, total n-6 and n-6 long-chain PUFA are higher at 5 years (Guerra et al., 2007).

Bjerve with coworkers (1989) studied age-related changes in the fatty acids of 735 subjects 12 to 89 years old. They report that the levels of 16:0, 16:0, and 18:2n-6 decrease from the third decade of life, the levels of 20:3n-6 and 20:4n-6 are invariable in all studied age periods, whereas the concentrations of long-chain n-3 PUFAs increase from the third to the fifth decade and remain invariable in old age. Authors suggest that old age per se adversely affects the n-3 PUFA status and that environment and food habits have even more considerable impact on the concentration of n-3 PUFAs in old people (Bjerve et al, 1989). The percentage of EPA + DHA in red blood cells, also known as the Omega-3 Index, which was proposed as a risk marker for death from coronary heart disease, is significantly depend on the age along with diabetes and body mass index; and the index increased by 0.5 units for each additional decade in age (Sands et al., 2005).

Age-related differences are documented for in the fatty acid composition of human adipose tissue (Boulton-Smith et al., 1997). The increase of 18:2n-6/18:3n-6 ratio, the contents of 20:3n-6 and sum of 22:5+22:6n-3 fatty acids with age do not depend on food habits. Nevertheless, the increase of 18:3n-6 content in a diet with age in order to maintain balance between PUFA levels in tissues is regarded as healthy.

Table 3. Fatty acid composition (%) of erythrocyte membranes healthy subjects of different age- and gender groups

Fatty acid	Boys n=18	Girls n=20	Men n=33	Women n=30
14:0	0.43±0.02	0.41 ±0.02	***0.68±0.03	0.58±0.15
14:1	–	–	0.20±0.02	0.15v0.07
a15:0	–	–	0.34±0.04***	0.10±0.01
15:0	0.27±0.01**	0.24±0.01	***0.41±0.02	0.35±0.06
15:1	–	–	0.07±0.01	0.1 ±0.01
i16:0	–	–	1.26±0.17***	0.07±0.03
16:0	21.76±0.50	20.80±0.20	***19.26±0.26***	21.10±0.26
16:1n-9	0.76±0.04	0.70±0.03	***1.32±0.10*	0.90±0.18
i17:0	–	–	0.23±0.04	0.18±0.05
a17:0	–	–	0.46±0.08***	0.1 ±0.01
17:0	0.43±0.03	0.39±0.01	**0.62±0.05***	0.40±0.01
17:ln-9	0.37±0.05	0.30±0.04	**0.21±0.03	*0.20±0.01
i18:0	–	–	1.84±0.23***	0.40±0.01
18:0	16.29+0.40	17.14±0.15	15.16±0.26*	16.55±0.55
18:ln-9	14.25±0.30	14.30 ±0.30	14.19±0.25	14.18±0.59
18:ln-7	–	–	1.40±0.12	1.35±0.26
18:ln-5	0.34±0.02	0.30±0.01	**0.20±0.04	0.23±0.06
18:2n-6	14.60±0.40*	13.50±0.20	***11.01±0.26	12.70±1.17
18:3n-6	–	–	0.04±0.01***	0.30±0.01
18:3n-3	0.17±0.02	0.16±0.01	*0.11±0.02*	0.23±0.05
18:4n-3	0.16±0.01*	0.13± 0.01	**0.55±0.11**	0.15±0.07
20:0	0.10±0.01	0.10±0.01	**0.30±0.07	0.13±0.05
20:1	0.3 5±0.04	0.33± 0.01	0.30±0.03	0.33±0.12
20:2n-6	–	–	0.67±0.08	0.43±0.22
20:3n-6	1.40±0.06*	1.20±0.05	***1.05±0.01	1.25±0.13
20:3n-3	–	–	0.10±0.05	0.18±0.09
20:4n-6	12.90±0.27*	13.80±0.25	***10.81±0.45	12.20±1.45
20:5n-3	0.84±0.08	0.90±0.05	***! 55±0 14***	0.87±0.07
22:4n-6	2.48±0.07	2.70±0.10	**1.98±0.16	1.93±0.90
22:5n-6	0.30±0.03	0.30±0.01	0.28±0.08	0.23±0.05
22:5n-3	2.04±0.09	2.04±0.06	2.15±0.07	2.14±0.30
22:6n-3	5.67±0.30*	6.56±0.20	***8.32±0.40	7.48±0.62
Σ n-3	8.27±0.08*	7.98±0.11	** 13.16±1.42	*11.92±1.75
Σ n-6	28.12±0.41	27.63±0.63	**25.96±0.55	28.40±2.76
Σ n-3/Σ n-6	0.30±0.01	0.29±0.01	***0.57±0.03	0.45±0.13

Note: Asterisks at the right indicate significant differences of fatty acid contents between men and women; asterisk at the left indicate significant differences of fatty acid contents between age groups: ** for $p < 0.01$; *** for $p < 0.001$.

The works mentioned above consider quantitatively differences in the fatty acid composition of blood serum and plasma lipids. Investigations of the fatty acids of erythrocyte membrane lipids in healthy subjects of different ages and genders are scarce.

We reported the results of the studies on age- and gender-related differences in the fatty acid composition of erythrocyte lipids in a group of healthy subjects from the Russian Far East (Vostrikova, Slezka, 1999; Endakova et al., 2002; Novgorodtseva et al., 2002). These results revealed low levels of saturated 14:0, 15:0, 17:0 and high concentrations of 16:0 and 18:0 fatty acids in boys compared to adult men (Table 3). The similar tendency for 14:0, 15:0, and 16:0 was distinguished between girls and women.

Age-related differences were more pronounced for n-6 and n-3 PUFA and were commonly detected in the male cohort of healthy subjects. The concentration of the n-6 fatty acids, namely 18:2n-6, 20:4n-6, 22:4n-6, and 22:5n-6, was higher in children, whereas the concentration of n-3 fatty acids, 18:4n-3, eicosapentaenoic acid (EPA) 20:5n-3, 22:5n-3, and particularly 22:6n-3, was higher in adults. Hence, the total n-3 PUFA content was lower and the total n-6 PUFA content was higher in children than in adults.

Table 4. The ratio of polyunsaturated fatty acids in erythrocyte membranes as indices of metabolic reactions for healthy subjects of different age- gender groups

Ratio	Boys, n=18	Girls, n=20	Men, n=33	Women, n=30
18:2n-6 / 20:4n-6	1.13 ± 0.33	0.98 ± 0.04	1.07 ± 0.16	1.04 ± 0.33
20:4n-6 / 20:5n-3	10.41 ± 0.49*	15.30 ± 2.15	9.91 ± 0.91	**7.60 ± 1.6 1
18:4n-3 / 18:3n-3	0.94±0.01***	0.81 ± 0.02	**2.14 ± 0.35*	*0.65 ± 0.06
20:4n-6 / 20:3n-6	9.23±0.17***	11.50 ± 0.24	* 10.56 ± 0.61	10.13 ± 1.77
22:6n-3 / 22:5n-3	2.98 ± 0.11	3.21 ± 0.20	***4.12 ± 0.20	3.50 ±0.59
20:4n-6 / 22:6n-3	2.32 ± 0.09	2.10 ± 0.27	*** 1.16 ± 0.06	1.65 ± 0.34
(20:3 ± 20:5) / 22:6	0.36 ± 0.02	0.32 ± 0.05	*0.28 ± 0.03	0.40 ± 0.08

Note: Asterisks at the right indicate significant differences of fatty acid contents between men and women; asterisks at the left indicate significant differences of fatty acid contents between age groups: * for $p< 0,05$; ** for $p<0.01$; *** for $p<0.001$.

The fatty acid composition of erythrocyte phospholipids reflects the endogenous turnover. We suppose that the variations in the composition of the PUFA composition may be caused by different rates of fatty acid metabolism in children and in adults (Table 4). Children had higher values of 20:4n-3/22:6n-3 and 20:4n-6/20:5n-3 ratios. The low value of 18:4n-3/18:3n-3 ratio in boys versus adult men was due to a very low level of 18:4n-3 in children. Additionally, boys had a high ratio of 20:3n-6+20:5n-3 to 22:6n-3.

The most characteristic feature of the fatty acid metabolism was the low value of 22:6n-3/22:5n-3 ratio in boys as compared to adult men. The ratios between PUFA suggest different activity of enzymes involved in fatty acid metabolism in children and adults.

Gender-related differences generally concerned the concentrations of n-6 and n-3 PUFA. For instance, female subjects had lower level of 18:4n-3 than males and this difference was more pronounced in adults than in children. The concentration of 22:6n-3 in erythrocyte membranes was higher in girls than in boys. Within the n-6 acid family, the levels of 18:2n-6 and 20:3n-6 fatty acids were significantly higher in boys, whereas the level of 20:4n-6 was higher in female healthy subjects than in their male counterparts. Male and female children differed in the total n-6 PUFA. Adult men showed an obvious decrease in the total n-6 fatty acids compared to women. Gender-related variations of fatty acid metabolism values were mostly pronounced in children (Table 4). Significant differences between boys and girls were revealed for the ratios of 18:4n-3/18:3n-3 and 20:4n-6/20:3n-6; the ratio of 18:4n-3 to 18:3n-3 was significantly higher and the ratio of 20:4n-6 to 20:3n-6 was lower in boys than in girls. The ratio of 22:6n-3 to 22:5n-3 had tendency to reduction in boys compared with the girls. These ratios characterizing the activity of fatty acid metabolism enzymes have a high diagnostic importance. The deficiency in n-3 PUFA caused by the suppression of desaturase activity is associated with the risk of atherosclerosis development. The low level of n-3 PUFA in boys shows a risk for ischemic heart disease. Atherosclerosis is known to begin in childhood, and early initiation of prevention measures through the diagnostic and behavioral means may lower the risk of future cardiovascular disease (Zappalla, Gidding, 2009).

Age- and gender-related differences in fatty acid composition were considered in the papers cited above (Bjerve et al, 1989; Decsi, Koletzko, 1994; Al et al., 1996; Boulton-Smith et al., 1997; Sumner et al., 2002; Guerra et al., 2007; Magkos et al. 2009). However these papers deal with mainly the fatty acid differences of serum and plasma lipid fractions. Investigation of the fatty acids of erythrocytes of the healthy subjects of different age and sex related groups are scare (Jakobik et al., 2009). The values of the fatty acid composition obtained in our study may serve as reference values of the fatty acid composition of erythrocyte membrane lipids for Russian Far East region.

Fatty Acids of Erythrocyte Membrane and the Diagnosis of Cardiovascular Disease

Cardiovascular disease (CVD) is the leading cause of mortality in developed countries. One of the most urgent problems in modern cardiology at present is the development of CVD in young people. Disturbances of lipid metabolism, dislipidemia, mainly the elevated total cholesterol, TAG and atherogenic lipoproteins (such as LDL) are established as important risk factors for atherosclerosis and CVD. Additional blood lipid markers used as risk factors for CVD are the alterations of phospholipid and fatty acid compositions of erythrocyte membranes. There is a considerable amount of literature data documenting the alterations in fatty acid composition of different blood fractions (serum and plasma) in CVD patients (Decsi, Koletzko, 1994; Dewailly et al., 2001; Albert et al., 2002; Harris, von Schacky 2004; Simopoulos, 2008). Traditionally, the authors focuse on n-6 and n-3 fatty acids and their diagnostic importance, but an increasing number of recent publications study alterations in the membrane phospholipid and fatty acid composition of various blood cells and use the erythrocyte as a model of biological membranes (Lemaitre et al., 2002; Harris et al., 2004; Sun et al., 2007; Block et al., 2008; Harris, 2008; Jakobik et al., 2009). The main fatty acids of erythrocyte phospholipids mirror the fatty acid composition of the phospholipids of other cells and tissues of the body (Hoffman D.R. et al, 1993; Nettleton, 1995; Harris et al., 2004; Levant et al., 2007).

Our investigations showed that the initial alterations in the phospholipid and fatty acid composition of erythrocyte membranes are sensitive and reliable biochemical indicators of ischemic heart disease, which can be registered far earlier than alterations in plasma lipoproteins (Novgorodtseva T, 1994; Novgorodtseva T, 1998; Novgorodtseva T et al., 2002).

To find out the trends of changes in erythrocyte lipids we examined 48 ischemic heart disease patients. The patients involved in the investigation had obvious clinical manifestations of ischemic heart disease: angina pectoris, myocardial infarction, and atherosclerotic cardiosclerosis. Sixty-three healthy subjects were selected as a control group to obtain the normal values for erythrocyte fatty acid composition, which were compared with the fatty acid values in CVD patients to estimate the possible alterations. The groups were matched by age and gender.

The fatty acid composition of erythrocyte membranes in the CVD patients exhibited the alteration in the content of saturated, monounsaturated fatty acids, and PUFAs compared with the control healthy group; the proportion of unsaturated fatty acids decreased (Tables 5 and 6). The most pronounced changes in CVD patients were found in the saturated fatty acid content caused by the increase in palmitic acid (16:0) and the decrease in margaric acid (17:0). The content of monounsaturated fatty acids altered insignificantly because of the increase in 17:1, 20:1, and 18:1n-9. For n-6 PUFA, the increase in linoleic acid (18:2n-6) was revealed, the estimates reaching 14.88 ± 0.49 in 47.8% of women and 12.74 ± 0.27 in 65.7% of men (Table 5). Among the changes in C20 n-6 fatty acids, we registered an increase in 20:3n-6 and a tendency to increase in 20:4n-6 concentration. Compared with healthy subjects, there was an elevated concentration of 20:4n-6 (up to 12.19%) in 68.6% of the examined men; in contrast, 65.2% of the women exhibited a relatively low level of this fatty acid. The concentration of 22:4n-6 fatty acid in 57.1% of men was as high as 2.18 ± 0.10 and reached 2.33 ± 0.06 in 43.5% of women. A decrease in 18:4n-3 and 22:6n-3 contents was responsible for the alteration of the total n-3 PUFA in the erythrocyte membranes. A similar pattern was found both in male and female cardiac patients. The concentration of 20:5n-3 lowered to 0.97 ± 0.04 in 34.3% of men and was within 1.08 ± 0.11 in 60.9% of women. The level of docosapentaenoic acid (22:5n-3) was in most CVD patients lower than in healthy subjects (1.84 ± 0.15 in 57.1% of men and 1.97 ± 0.14 in 87% of women).

The similar results were reported by Lausada et al., 2007. Plasma and erythrocyte membrane from coronary heart disease patients compared to controls, showed a decrease in polyunsaturated fatty acids and an increase in saturated fatty acids leading to a decrease in the unsaturation index (plasma: 1.67 vs. 1.28; erythrocyte membrane: 2.28 vs. 1.25) and an enhancement in the saturated/unsaturated ratio (plasma: 0.35 vs. 0.52 erythrocyte membrane: 0.45 vs. 0.83). These data are consistent with an essential fatty acid deficiency. Total cholesterol was increased in the erythrocyte membrane of the coronary heart disease patients (32.3 vs. 40.6) with a decrease in phospholipid percentage (67.7 vs. 59.4) indicating an alteration in membrane fluidity.

The alterations of fatty acid composition revealed in this study are similar to those known for cardiac pathologies from other publications. Decsi T. and Koletzko B. (1994), Albert et al. (2002), Harris and von Schacky (2004), Block et al., (2008) have demonstrated a considerable increase in the proportion of saturated fatty acids and a decrease in n-3 PUFA content in patients with ischemic heart disease, myocardial infarction, and other diseases characterized by dislipidemia. The deficiency of essential fatty acids, mainly linoleic acid, is an important risk factor for CVD (Decsi T., Koletzko B., 1994). We did not find a

Table 5. Fatty acid composition of erythrocyte membranes of patients with CVD, mean ± SE

Fatty acid	Healthy subjects Men n=33	Healthy subjects Women n=30	Patients with CVD Men n=35	Patients with CVD Women n=23
14:0	0.68 ± 0.03	0.58±0.15	0.71±0.05	0.70±0.04
14:1	0.20 ±0.02	0.15±0.07	0.16±0.02	0.15±0.01
a15:0	0.34±0.04	0.10±0.01	0.16±0.02	0.14±0.02
15:0	0.41±0.02	0.35± 0.06	**0.30±0.03	0.40±0.03
15:1	0.07±0.01	0.10±0.01	0.18±0.02	0.22±0.05
i16:0	1.26±0.17	0.07±0.03	***0.13±0.02	0.12±0.01
16:0	19.26± 0.26	21.10± 0.26	***23.35±0.58	***24.12±0.62
16:1n-9/7	1.32±0.10	0.90±0.18	***1.28±0.10	1.30±0.08
i17:0	0.23±0.04	0.18±0.05	0.21±0.10	0.21±0.01
a17:0	0.46±0.08	0.10±0.01	**0.16±0.01	0.16±0.02
17:1n-9	0.21±0.03	0.20± 0.01	**0.36±0.04	**0.41±0.07
i18:0	1.84±0.23	0.40±0.01	***0.14±0.02	***0.13±0.02
18:0	15.16± 0.26	16.55± 0.55	15.88±0.34	15.80±0.42
18:1n-9	14.19± 0.25	14.18± 0.59	*15.49±0.43	15.44±0.46
17:0	0.62± 0.05	0.40± 0.01	*0.47±0.01	0.49±0.03
18:1n-7	1.40±0.12	1.35±0.26	1.67±0.05	1.61±0.06
18:1n-5	0.20±0.04	0.23± 0.06	0.19±0.02	0.22±0.01
18:2n-6	11.01± 0.26	12.70±1.17	*12.01±0.30	12.53±0.39
18:3n-6	0.04±0.01	0.30±0.01	0.30±0.01	0.20±0.02
18:3n-3	0.11± 0.02	0.23± 0.05	***0.20±0.01	0.21±0.01
18:4n-3	0.55± 0.11	0.15± 0.07	***0.16±0.03	0.17±0.03
20:0	0.30±0.07	0.13± 0.05	***0.13±0.01	0.16±0.01
20:1	0.30±0.03	0.33± 0.12	*0.40±0.02	0.40±0.02
20:2n-6	0.67±0.08	0.43±0.22	0.81±0.13	0.68±0.12
20:3n-6	1.05± 0.01	1.25± 0.13	***1.38±0.07	1.37±0.02
20:3n-3	0.10±0.05	0.18±0.09	0.13±0.02	0.26±0.08
20:4n-6	10.81± 0.45	12.20±1.45	11.19±0.49	10.68±0.58
20:5n-3	1.55± 0.14	0.87± 0.07	1.46±0.12	1.37±0.12
22:4n-6	1.98± 0.16	1.93± 0.90	1.82±0.10	1.77±0.12
22:5n-6	0.28± 0.08	0.23± 0.05	0.21±0.02	0.20±0.02
22:5n-3	2.15±0.07	2.14± 0.30	2.26±0.13	2.08±0.13
22:6n-3	8.32±0.40	7.48± 0.62	***5.84±0.35	**5.46±0.38
Σn-3	13.16±1.42	11.92±1.75	***9.80±0.52	9.69±0.65
Σn-6	25.96±0.55	28.40±2.76	27.42±0.74	28.26±1.00

Note: Asterisks indicate significant differences of fatty acid contents between healthy and CVD patients: * for $p < 0.05$; ** for $p < 0.01$; *** for $p < 0.001$.

Table 6. Specificity of distribution of the main polyunsaturated fatty acids (% of the total fatty acids) in erythrocyte membranes of patients with CVD with respect to the normal values for healthy subjects, mean ± SE

Fatty acid	Deviation from the norm	*Patients with CVD*	
		Men n=35	Women n=23
18:2n-6	E[§]	12.74±0.27*** n=23 (65.7%)	14.88±0.49 n=11 (47.8%)
	M	12.01±0.30*	12.53±0.39
	R	10.16±0.15** n=6 (17.1%)	11.15±0.36 n=7 (30.4%)
18:3n-3	E	0.23±0.01*** n=29 (82.9%)	0.30±0.05 n=3 (13%)
	M	0.20±0.01***	0.21±0.01
	R	–	0.18±0.01 n=20 (86.9%)
20:4n-6	E	12.19±0.25* n=24 (68.6%)	13.83±0.84 n=3 (13%)
	M	11.19±0.49	10.68±0.58
	R	9.02±0.44** n=7 (20%)	9.32±0.75 n=15 (65.2%)
20:5n-3	E	2.07±0.16* n=15 (42.9%)	–
	M	1.46±0.12	1.37±0.12***
	R	0.97±0.04*** n=12 (34.3%)	1.08±0.11 n=14 (60.9%)
22:4n-6	E	2.18±0.10 n=20 (57.1%)	2.33±0.06 n=10 (43.5%)
	M	1.82±0.10	1.77±0.12
	R	1.37±0.08** n=12 (34.3%)	1.36±0.15 n=12 (52.2%)
22:5n-3	E	2.69±0.10*** n=13 (37.1%)	–
	M	2.26±0.13	2.08±013
	R	1.84±0.15 n=20 (57.1%)	1.97±0.14 n=20 (87%)
22:6n-3	E	–	8.30±0.28 n=2 (8.7%)
	M	5.84±0.35***	5.46±0.38***
	R	6.07±0.25*** n=35 (100%)	5.01±0.41** n=19 (82.6%)
∑n-3	E	14.60 n=1 (2.9%)	–
	M	9.80±0.52*	9.69±0.65
	R	10.11±0.35* n=34 (97.1%)	8.53±0.77 n=17 (73.9%)
∑n-6	E	28.47±0.42*** n=28 (80%)	31.82±0.27 n=7 (30.4%)
	M	27.42±0.74	28.26±1.00
	R	22.73±1.41* n= 5 (14.3%)	24.03±1.85 n=8 (34.8%)
n-3/n-6	E	–	0.50±0.06 n=1 (4.3%)
	M	0.30±0.04***	0.34±0.02
	R	0.37±0.01*** n=34 (97.1%)	0.29±0.02 n=17 (73.9%)
IU	E	178.14±2.30 n=8 (22.9%)	181.60±3.33 n=3 (13%)
	M	158.68±4.01	154.53±5.98
	R	151.75±2.75 n=25 (71.4%)	148.09±6.97 n=19 (82.6%)

Note: Asterisks indicate significant differences of fatty acid contents between patients with CVD and healthy subjects: * for $p < 0,05$; ** for $p < 0.01$; *** for $p < 0.001$.

[§] E – elevated, M – middle, R – reduced levels.

significant decrease in the levels of linoleic and α-linolenic acids. On the contrary, male and female CVD patients in our research had raised concentrations of LA, and only in 17.1% of men and 30.4% of women had a low level of LA. The ALA level was actually reduced only in women (85.9%).

The evidence of a role of essential fatty acids and their metabolites in the pathobiology of coronary heart disease, diabetes mellitus and essential hypertension gave by Das (1995). Fatty acid analysis of the plasma phospholipid fraction revealed that in coronary heart disease the levels of γ-LA, AA, EPA and DHA are low compared to normal control. These results are interesting since DGLA, AA and EPA form precursors to prostaglandin E_1, (PGE_1), prostacyclin (PGI_2), and PGI_3, which are potent platelet anti-aggregators and vasodilators and can prevent thrombosis and atherosclerosis. These results suggest that an alteration in the metabolism of essential fatty acids is closely associated with coronary heart disease. Since features of coronary heart disease are associated with decreased concentrations of PUFA in the plasma, the possibility is raised that changes in the metabolism of essential fatty acids may have a fundamental role in the pathobiology of these conditions.

The correlation between cardiovascular diseases and changes in the blood fatty acid composition has been recently investigated by methods of population studies. Additional diagnostic opportunities and informative importance of erythrocyte fatty acid composition has been proved in the present study by statistical analysis (Novgorodtseva et al., 2001).

A statistical analysis was performed for 128 subjects with the use of the data on their anthropometric characteristics, heredity, medical history, clinical and biochemical examinations, and erythrocyte fatty acid composition. All members of the study group were residents of the city of Vladivostok older than 18 years of age and living in this climatic zone for more than 10 years. Based on the clinical data and medical history, the examined subjects were divided into the following groups: (1) 26 healthy subjects; (2) 15 tentatively (nominally) healthy subjects with a family history of atherosclerosis and CVD; (3) 30 patients with neurocirculatory dystonia; (4) 17 patients with arterial hypertension; (5) 40 patients with angina pectoris and myocardial infarction.

A statistical analysis of our results was performed using software the STATISTICA 6.0 for WINDOWS. The fatty acid values of the subjects examined were compared with disease severity (5 groups) by the use of one-way analysis of variance (ANOVA) followed by Fisher's test and Bartlett's test. The ANOVA results are given in Table 7. Fisher's test and Bartlett's test showed significant differences between the majority of the fatty acids of erythrocyte membranes and disease severity in patients with CVD. These results confirm that the values of some fatty acids and fatty acid ratios are suitable for use as diagnostic criteria of CVD.

Table. 7. The Bartlet's test for fatty acids of erythrocyte membranes of CVD patients

Fatty acid	Bartlet's test	p
14:0	$B=1.52$	$p=0.18$
a15:0	$B=11.98$	$p<<0.01$*
15:0	$B=1.28$	$p=0.3$
14:1	$B=4.50$	$p<<0.01$
15:1	$B=4.08$	$p=0.001$*
16:0	$B=85.69$	$p<<0.001$*
16:0	$B=10.00$	$p<<0.01$*
16:1n-9/7	$B=1.13$	$p=0.35$
i17:0	$B= 1.21$	$p=0.3$
a17:0	$B=13.87$	$p<<0.01$*
17:0	$B=0.53$	$p=0.75$
17:1	$B=4.12$	$p<<0.01$*
i18:0	$B=94.19$	$p<<0.01$*
18:0	$B=2.66$	$p=0.02$*
18:1n-9	$B=1.76$	$p=0.12$
18:1n-7	$B= 85.60$	$p<<0.01$*
18:1n-5	$B=8.06$	$p<<0.01$*
18:2n-6	$B=4.57$	$p<<0.01$*
18:3n-6	$B=4.94$	$p<<0.01$*
18:3n-3	$B=4.17$	$p<<0.01$*
18:4n-3	$B=16.85$	$p<<0.01$*
20:0	$B=1.29$	$p=0.3$
20:1	$B=8.63$	$p<<0.01$*
20:2n-6	$B= 2.87$	$p=0.02$*
20:3n-6	$B=5.74$	$p<<0.01$*
20:4n-6	$B=2.40$	$p=0.05$*
20:3n-3	$B=2.20$	$p=0.05$*
20:5n-3	$B=2.45$	$p=0.04$*
22:4n-6	$B=2.35$	$p=0.05$*
22:5n-6	$B=2.54$	$p=0.04$*
22:5n-3	$B=2.34$	$p=0.05$*
22:6n-3	$B=16.74$	$p<<0.01$*
Σn-3	$B=11.59$	$p<<0.01$*
Σn-6	$B=4.31$	$p=0.001$*
ΣC_{20}-C_{22}n-3	$B=9.06$	$p<<0.01$*
ΣC_{20}-C_{22}n-6	$B=2.35$	$p=0.05$*
Fatty acid	*Bartlet's test*	p
Σ n-3/Σ n-6	$B=20.22$	$p<<0.01$*
(ΣC20-C22n-3) / (ΣC20-C22n-6)	$B=16.66$	$p<<0.01$*
18:4n-3 / 18:3n-3	$B=16.31$	$p<<0.01$*
20:4n-6 / 20:3n-6	$B=4.47$	$p<<0.01$*
22:6n-3 / 22:5n-3	$B=26.03$	$p<<0.01$*
20:4n-6 / 22:6n-3	$B=11.57$	$p<<0.01$*
(20:2n-6+20:5n-3) / 22:6n-3	$B=4.29$	$p=0.001$*
20:4n-6 / 20:5n-3	$B=2.41$	$p=0.05$*
20:4n-6 / 22:4n-6	$B=1.93$	$p=0.14$
20:5n-3 / 22:5n-3	$B=2.41$	$p=0.05$*
UI	$B=2.64$	$p=0.03$*

Note * indicates the significant differences

Associations between cardiovascular disease severity and the fatty acid composition of erythrocyte membranes of patients were examined by canonical correlation analysis. *The canonical correlation analysis enables to determine the associations between vectors: X=(x1, x2, …, xn) and Y=(y1, y2, …, ym).* Vector *Y* in our research characterized the disease severity of the patients. All patients were ranked according to their disease severity using a 12-point scale: 1 point for healthy subjects; 3 points for nominally healthy ones; 5 points for patients with neurocirculatory dystonia, which is a factor of CVD; from 6.5 to 8 points for patients with arterial hypertension, from 9.5 to 12 for patients with ischaemic heart disease, myocardial infarction, and angina pectoris. Vector *X* characterized the fatty acid compositions and the ratios of PUFAs.

The values of fatty acids were divided into the following groups: the concentrations of the main fatty acids (33 items in total); saturated fatty acids; monounsaturated fatty acids; fatty acids with chain lengths from C18 to C22; C18–C22 n-6 fatty acids; C18–C22 n-3 fatty acids; C20 n-6 and n-3 fatty acids; C18–C22 fatty acids; C20–C22 n-6 fatty acids; C20–C22 n-3 fatty acids; and 17 PUFA ratios as indicators of fatty acid metabolism (Table 8). The canonical correlation coefficients were calculated to evaluate the correlations between each group of fatty acid values and disease severity.

High values of the correlation coefficient (CR=0.71) for the first group prove the association of fatty acid concentrations with disease severity. The most significant associations with disease severity were detected for fatty acids with the number of carbon atoms from 18 to 22 (CR=0.67 for group 4, CR=0.66 for group 5, and CR=0.60 for group 6) suggesting that it is these fatty acids that play an important part in the pathologic process. The data in Table 8 show that simultaneous changes in the levels of n-3 and n-6 fatty acids are closer associated with disease severity (CR=0.60 for group 6 and CR=0.57 for group 9) than individual changes in the levels of each fatty acid family (CR=0.50; CR=0.51; CR=0.47; and CR=0.43 respectively for groups 7, 8, 10, and 11). Additionally, the correlation coefficient is rather high (CR=0.63) for the group with PUFA ratios as indicators of metabolism conversions.

This approach enabled us to determine relationships between the fatty acid composition and the CVD disease severity, as well as to differentiate the values of fatty acids that can be used as diagnostic indicators.

Table 8. The canonical correlation coefficients for the groups of fatty acids of erythrocyte membranes and degree of disease severity in patients with CDV

N	Group of the fatty acids	Number of fatty acids in group	CR
	Total fatty acids	33	0.71
	Saturated	11	0.62
	Monounsaturated	8	0.60
	C_{18}-C_{22}	20	0.67
	C_{18}-C_{22} PUFA	18	0.66
	C_{18}-C_{22} n-3, n-6	13	0.60
	C_{18}-C_{22} n-6	7	0.50
	C_{18}-C_{22} n-3	6	0.51
	C_{20}-C_{22} n-3, n-6	9	0.57
	C_{20}-C_{22} n-6	5	0.47
	C_{20}-C_{22} n-3	4	0.43
	C_{18} n-3, n-6	4	0.46
	C_{20} n-3, n-6	5	0.41
	C_{22} n-3, n-6	4	0.44
	Indices of metabolic conversation	17	0.63

The results obtained in this study suggest the following criteria for use in diagnostics:

- $\sum$n-3 and $\sum$n-6 values reflect the degree of unsaturation of the erythrocyte membrane. The decrease in $\sum$n-3 and increase in $\sum$n-6 promote the enhanced microviscosity of the cell membrane, which is a risk factor for CVD development.
- The fatty acid unsaturation index, calculated as the sum of double bond productions in each fatty acid divided by its percentage of the total fatty acids, accounts for the overall changes in the fatty acid composition in total and characterizes the degree of membrane unsaturation.
- The ratios of n-6 to n-3 PUFAs reflect the disorders in the eicosanoid cycle. These are the ratio of eicosanoid synthesis precursor to the inhibitor of this synthesis (20:4n-6/22:6n-3); the ratio of the total precursors of eicosanoid synthesis to its inhibitor ([20:2n-6+20:5n-3]/ 22:6n-3); and $\sum$n-3/$\sum$n-6 and ($\sum C_{20}$-C_{22} n-3)/($\sum C_{20}$-C_{22}n-6) ratios. The increase in the first two criteria and decrease in the latter two results from blood clot organization, which frequently attends CVD development.
- The ratios between PUFAs characterize the activity of desaturases and elongases. The decrease in 22:6n-3/22:5n-3 ratio points to the suppression of enzymes at the last step of fatty acid synthesis, common in

CVD patients. This value is most informative at the early stages of the disease. The increase in 20:4n-6/22:4n-6 ratio indicates the possible decrease in the activity of fatty acid elongation, which always accompanies the development and progress of CVD.

To sum up, the analysis of variance and correlation analysis used in this study confirmed the great diagnostic importance of the fatty acids of erythrocyte lipids in CVD development, which had been shown also by previous medical and biological investigations.

Specific Blood Lipid Alterations in Hepatobiliary Diseases

The abnormalities in the phospholipid and fatty acid composition in erythrocyte membranes and plasma are recognized as a key to the diagnosis of the some therapeutic diseases. Liver function influences the homeostasis of complex lipids in plasma (Arranz et al., 1996). The hepatobiliary system diseases can cause metabolic and inflammatory alterations. Liver disease is accompanied by major quantitative and qualitative modifications in plasma lipoprotein metabolism (Taus et al., 1999). The liver plays a central role in the metabolism of polyunsaturated fatty acids. Patients with chronic liver diseases are at a high risk of essential fatty acid deficiency which is correlated with progressive elevation of serum bilirubin and progressive deterioration of oxidant status (Abdel-Ghaffar et al., 2003).

The abnormal fatty acid composition of blood is described for hepatobiliary diseases by several authors (Robberecht et al., 1997; Socha et al., 1997a, b). The content of essential fatty acids in phospholipids of blood serum of children suffering of cholestasis changed insignificantly, but the development of the disease was accompanied by deficit of their long-chain metabolites. The level of bilirubin correlated at cholestasis with the contents of fatty acids 20:4, 22:5, and 22:6 (Socha et al., 1997a). Similar regularities were investigated at hyper-bilirubinemia (Robberecht et al., 1997). These results are quite explicable, since the liver plays a central role in the long-chain fatty acids exchange and in the metabolism of polyunsaturated fatty acids. Correspondingly, liver dysfunction leads to defects in fatty acid metabolism. Patients with chronic liver diseases are at a high risk of essential fatty acid deficiency which is correlated with progressive elevation of serum bilirubin and progressive deterioration of oxidant status (Abdel-Ghaffar et al., 2003).

The abnormal fatty acid composition of serum phospholipids and cholesterol esters was revealed in hepatobiliary disease in a group of cholestatic children with extrahepatic biliary atresia. Relationships between fatty acids in phospholipids and cholesterol esters which exist in the control health group were either absent or different in the patient group. (Robberecht et al., 1997)

Socha with coworkers (1997a) defined the relationship between the essential fatty acid (EFA) status and indicators of liver function in children with chronic cholestasis. The cholestatic patients had very low values of phospholipid essential fatty acids, particularly linoleic acid (18:2 n-6) and its major metabolite arachidonic acid (20:4 n-6). The patients' essential fatty acid depletion was reflected by increased levels of the non-essential fatty acids, eicosatrienoic acid 20:3 n-9 and palmitoleic acid 16:1 n-7. Fatty acid values exhibited an inverse relationship to serum bile acids, as well as to serum bilirubin. Bilirubin values were unrelated to the essential fatty acid precursors linoleic acid and α-linolenic acid but correlated inversely with their long-chain metabolites, arachidonic acid, docosapentaenoic acid (22:5 n-3), and docosahexaenoic acid (22:6 n-3). Hepatic conversion of essential precursor fatty acids into their long-chain metabolites may be increasingly impaired with advancing severity of liver disease. Polyunsaturated fatty acid profiles did not differ between infants with biliary atresia and those with intrahepatic cholestasis, or between infants with severe malnutrition and mild malnutrition (Socha et al., 1998). Breastfed cholestatic infants had higher values of the n-3 long-PUFA, docosapentanoic acid (22:5 n-3) and docosahexanoic acid (22:6 n-3), than formula-fed infants did, while disease severity was similar in the two groups. Plasma lipid peroxides were significantly higher in cholestatic patients than in controls and correlated with total bilirubin. Thus cholestatic infants are at a high risk of essential fatty acid depletion, which appears to be related to fat malabsorption, hepatic essential fatty metabolism, enhanced lipid peroxidation and dietary intake.

Oral supplementation of vitamin E, which plays an important role as a free-radical scavenger, to cholestatic children does not improve the increased lipid peroxidation and poor polyunsaturated fatty acid status (Socha et al., 1997b). Long-chain PUFA supplemented formulae improve the long-chain status of infants with severe cholestasis but may enhance lipid peroxidation (Socha et al., 2002).

A great deal of new information has been obtained on determination of disturbances in the lipid composition at different hepatobiliary diseases. Many studies revealed changes in the fatty acid content and disturbances of lipid metabolism but effect of the diseases on PUFA status and lipid metabolism are contradictory. The aims of our study were to determine serum lipids as well as the phospholipid composition and the essential fatty acid status in erythrocytes of patients with hepatobiliary diseases and to study the relationship between the lipid characteristics, the antioxidant status and disease severity.

A group of 297 patients with hepatobiliary (HB) pathology were investigated. The group included men and women with different diseases of hepatobiliary system, namely biliary dyskinesia (55 subjects), cholelithiasis (38 subjects), and chronic acalculous cholecystitis (204 subjects). 40 healthy subjects as controls were selected to obtain the normal lipid pattern profile. A group of 85 patients with CVD was also studied. Alterations in the blood lipid composition were estimated with relation to a control group of healthy subjects and patients with CVD.

Blood biochemical assays were performed in the group of hepatobiliary patients. Measurements were taken of serum concentrations of total lipids, cholesterol, antioxidant activity and plasma lipid peroxidation expressed as level of malondialdehyde, as well as the phospholipid and fatty acid compositions of erythrocyte membranes. Table 9 shows that a majority of analyzed characters differed from corresponding values in the control group of healthy subjects. The high values of serum total cholesterol, LDL cholesterol, HDL cholesterol and apo B in patients with HB pathology exceeded those for the control healthy subjects.

The health risk associated with elevated plasma lipids is largely dependent on the dynamics of metabolic processes of the different major lipoprotein fractions (i.e. chylomicrons, VLDL, LDL and HDL) (Karpe, 1997; Knuth, Horowitz, 2006). Therefore, measurement of the level of different lipid fractions provides information on the atherogenic potential of the blood lipids.

The HDL cholesterol content appeared to be a more informative predictor of the coronary heart disease than the total cholesterol or LDL cholesterol contents. More illustrative are the indices presenting the ratios of various lipoproteins. It was proposed to evaluate the «predictor" as the so-called Index of atherogenesis (IA), which is calculated as the ratio of the sum of LDL cholesterol and VLDL cholesterol to HDL cholesterol (Klimov, Nagornev, 1994). To simplify the calculation of this index, we can admit that merely two measured values are sufficient: total serum total cholesterol and HDL cholesterol. In the case, the Index of Atherogenesis is determined as the ratio of the difference of the total cholesterol and HDL cholesterol to HDL cholesterol.

IA = (total serum cholesterol – HDL cholesterol) / HDL cholesterol.

Table 9. Blood serum lipids and antioxidant activity in the control and patient groups, mean ± SE

Parameters	Healthy subjects n=40	Patients with CVD (control group), n=85	Patients with HBD, n=297
Total cholesterol, mmol·L^{-1}	4.55 ± 0.07	5.71 ± 0.16**	5.23 ± 0.08***
ТГ, ммоль/л mmol·L^{-1}	1.31 ± 0.05	1.69 ± 0.10	1.26 ± 0.05
HDL cholesterol, mmol·L^{-1}	1.21 ± 0.03	1.16 ± 0.07	1.36 ± 0.03**
VLDL cholesterol, mmol·L^{-1}	0.59 ± 0.02	0.77 ± 0.05	0.57 ± 0.02
LDL cholesterol, mmol·L^{-1}	2.74 ± 0.09	3.65 ± 0.18***	3.22 ± 0.09**
Aterogenicy index	2.85 ± 0.11	5.05 ± 0.42***	3.4 ± 0.15**
Apo B	81.0 ± 1.19	-	107.08 ± 7.23**
MDA, nmol·g^{-1} Hb	8.3 ± 0.09	8.15 ± 0.27	7.23 ± 0.13
Antioxidative activity, %	54.18 ± 0.91	46.93 ± 3.30	53.44 ± 1.27

Note: Asterisks indicate significant differences of lipid fraction contents between the patients with HBD and control subjects: ** for $p < 0.01$; *** for $p < 0.001$

Naturally, the higher the index value is the greater is the danger of the occurrence and advance of the disease. Thus, the value of the IA is positively associated with advancement in the development of disease. Calculation of this index is widely used in the clinical practice in Russia because of its informative value and relative simplicity of estimation of total serum cholesterol and HDL cholesterol.

We found that the values of the index of atherogenesis were high in the patient groups compared to that of the healthy group. Thus, liver damage and alterations in the exocrine function of the gland lead to a profound alteration of the plasma lipoprotein profile.

The concentrations of malondialdehyde (MDA), an index of lipid peroxidation, and total antioxidant activity in the blood of the patients with HBD corresponded to the reference values (Table 9).

Table 10. Phospholipid composition of erythrocytes (%) in the subjects examined, mean ± SE

Phospholipids	Healthy Subjects, N=40	Patients With CVD (Control Group), N=36	Patients With HBD, N=35
Phosphatidylserine (PS)	10.47±0.07	9.57±0.39**	12.48±0.37***
Sphingomyelin (SM)	29.14±0.53	23.82±0.83***	25.48±0.53***
Phosphatidylcholine (PC)	35.04±0.09	34.44±0.68	32.61±0.49***
Phosphatidylethanolamine (PE)	25.34±0.55	32.13±0.85***	29.44±0.47***
PC/PE Ratio	1.41±0.03	1.08±0.02***	1.12±0.03***

Note: Asterisks indicate significant differences of phospholipid contents between the patients with HBD and control subjects: ** for $p < 0.01$; *** for $p < 0.001$

Two-dimensional TLC analysis of total phospholipids of erythrocytes of patients with hepatobiliary pathology showed the presence of six components distinguished on the chromatograms as areas of different intensities. Among phospholipid classes, phosphatidylcholine (PC), phosphatidylethanolamine (PE), sphingomyelin (SM), phosphatidylserine (PS), phosphatidylinositol (PI), and phosphatidic acid (PA) were identified. PC was predominant (Table 10). The next in quality were PE and SM. PI and PA were detected in trace amounts. Apparently, in the case of pathology, the redistribution of phospholipids occurred. The erythrocyte membrane phospholipids of patients with hepatobiliary pathology distinguished from those of the control group by their elevated concentrations of PE and PS and the reduced proportions of PC and SM.

The fatty acid composition of erythrocyte lipids was analyzed by capillary gas-liquid chromatography. We identified 48 components: fatty acids with chain length from C10 to C22, even and odd, straight and branched (iso-), saturated, mono- and polyunsaturated were identified. The dominant fatty acids (more than 5%) were 16:0, 18:0, 18:1n-9, 18:2n-6, 20:4n-6, and 22:6n-3. Table 11 summarizes the fatty acid composition of erythrocyte membranes in the case and the control subjects.

Fatty acid patterns divergent from the controls were described in patients with hepatobiliary diseases. Compared with the control subjects, HBD subjects had higher values of some saturated fatty acids, mainly 16:0, and the sum of n-6 PUFA and lower percentages of branched (iso- and anteiso-) fatty acids and the sum of n-3 PUFA, mainly at the expense of low relative DHA values. Increased plasma level of saturated fatty acids, decreased levels of n-3 and n-6 PUFA, and increased n-6/b-3 PUFA ratio were found at development of non-alcoholic steatohepatitis in the transgenic rats (Malinska et al., 2009). These results are consistent with the hypothesis that increased oxidative stress and disturbance in the metabolism of saturated fatty acids and PUFA may precede and contribute to the primary pathogenesis of hepatobiliary disease. Decrease in plasma phospholipid fatty acids was reported for patients with end-stage liver disease (Zeh et al., 2003). These phospholipid fatty acids have a lower proportion of the PUFA n-3 and n-6 with the result that the plasma level of these essential PUFA is extremely low compared to that of healthy subjects. Infants with HBD demonstrate sings of essential fatty acid deficiency, characterized by decreases AA 20:4n-6 and increased 16:1n-9 and 18:0 acids (Pettei et al., 1991). All of infants have substantial concentrations of the pathologic trienoic 20:3n-9 and reduced 18:2n-6 concentrations. Because the liver plays a central role in the metabolism of PUFA, hepatic conversation of essential precursor fatty acids into their long-chain metabolites may be increasingly impaired with advancing severity of liver disease (Socha et al., 1997).

Table 11. Fatty acid composition (%) of erythrocytes in patients and control groups, mean ± SE

Fatty acids	Healthy subjects n=40	CVD patients (control group) n=36	Patients with HBD, n=35
14:0	0.69 ± 0.02	0.67 ± 0.03	0.91 ± 0.06**
15:0a	0.24 ± 0.03	0.13 ± 0.01**	0.13 ± 0.01**
15:0	0.37 ± 0.01	0.36 ± 0.02*	0.39 ± 0.02
15:1	0.07 ± 0.005	0.14 ± 0.01***	0.16 ± 0.19
16:0i	1.18 ± 0.15	0.12 ± 0.01***	0.13 ± 0.03***
16:0	20.31 ± 0.25	23.52 ± 0.35***	22.22 ± 0.35***
16:1n-9	1.20 ± 0.07	1.19 ± 0.07	1.58 ± 0.1**
17:0i	0.22 ± 0.02	0.19 ± 0.01	0.19 ± 0.03
17:0a	0.37 ± 0.06	0.15 ± 0.012**	0.14 ± 0.01***
17:0	0.51 ± 0.03	1.43 ± 0.95	0.39 ± 0.01***
17:1	0.15 ± 0.01	0.33 ± 0.03***	0.31 ± 0.31
18:0i	1.77 ± 0.2	0.12 ± 0.008***	
18:0	15.71 ± 0.15	16.42 ± 0.24*	16.11 ± 0.31
18:1n-9	14.11 ± 0.21	15.17 ± 0.23***	15.34 ± 0.28***
18:1n-7	0.55 ± 0.11	1.61 ± 0.05***	2.12 ± 0.05***
18:2n-6	11.65 ± 0.22	12.29 ± 0.25	14.42 ± 0.52***
18:3n-6	0.15 ± 0.01	0.16 ± 0.02	0.31 ± 0.02***
18:3n-3	0.16 ± 0.01	0.20 ± 0.01*	0.31 ± 0.03***
18:4n-3	0.51 ± 0.08	0.11 ± 0.013***	0.21 ± 0.01***
20:2n-6	0.57 ± 0.06	0.61 ± 0.1	1.05 ± 0.12**
20:3n-6	1.13 ± 0.04	1.38 ± 0.05***	1.17 ± 0.04
20:4n-6	11.06 ± 0.22	11.25 ± 0.28	11.49 ± 0.36
20:5n-3	1.18 ± 0.10	1.27 ± 0.08	1.12 ± 0.07
22:4n-6	1.81 ± 0.07	1.88 ± 0.07	1.8 ± 0.08
22:5n-6	0.21 ± 0.03	0.13 ± 0.01*	0.58 ± 0.02
22:5n-3	2.28 ± 0.03	2.21 ± 0.06*	2.10 ± 0.07
22:6n-3	8.80 ± 0.29	5.64 ± 0.23***	5.36 ± 0.18***
Σ n-3	13.05 ± 0.30	9.55 ± 0.32***	8.9 ± 0.29***
Σ n-6	26.61 ± 0.47	27.72 ± 0.41	30.12 ± 0.56***
n-3/n-6	0.50 ± 0.01	0.36 ± 0.01***	0.29 ± 0.01***
IU	170.67 ± 1.54	155.54 ± 2.43***	157.38 ± 2.44***

Note: Other fatty acids include 10:0, 12:0, 19:0, 20:0, 22:0, 14:1, 18:1n-5, 20:1, 22:1, 18:2n-5, and 20:3n-3, which were present in trace amounts (<0.2%). Asterisks indicate significant differences of fatty acids contents between the patients with HBD and control subjects: * for $p < 0.05$; ** for $p < 0.01$; *** for $p < 0.001$.

Additionally, the HBD and CVD subjects also had a lower Unsaturation Index (UI) than the control group. A relative content of palmitic acid 16:0 and the sum of n-6 increased, and branched (iso- and anteiso-) fatty acids and the sum of n-3, mainly 22:6n-3 decreased, this resulting in a lower unsaturation index than control. The alterations of the fatty acid composition of erythrocytes were common for HBD and CVD patients. A lower unsaturation index compared to that in the control group and significant alterations of the fatty acid composition of erythrocytes were common for HBD and CVD patients. We suggested that in general, a low value of unsaturation index and the ratio of n-3 to n-6 suggest that increase in the packing density of the phospholipid molecules in the membrane causes changes in the structure of the blood cell membrane. The revealed modification of the structure of the phospholipid bilayer of the erythrocyte membrane contributes to the membrane decomposition, function abnormality of membrane, failure of its barrier and transport functions, damages in the transmembrane gradients.

With the progress of deterioration of hepatic function, the concentration of plasma n-3 fatty acids decrease in postoperative patients with biliary atresia (Takamatsu et al., 1999). The ratio of 20:3n-6 to 20:4n-6 is increased significantly with the progress of liver dysfunction. The activity of Δ5 desaturase is suspected to be suppressed in biliary atresia patients with poor liver function.

DHA deficiency would be secondary appeared to fat malabsorption. However the patients with extrahepatic biliary atresia, which receive emulsions, containing LA and ALA in addition to a diet, have lower percentage of DHA in both plasma and red blood cell phospholipids in comparison to the normal group (Yamashiro et al., 1994). This demonstrates a deficiency in the long-chain acid desaturase activity of these patients and disturbance in fatty acid metabolism.

Later the decrease in the n-3/n-6 ratio of PUFA was compared among the groups of patients with biliary atresia after (1) Kasai operation, (2) after liver transplantation, and (3) normal control group (Sumida et al., 2009). There is no significant difference between the transplantation and control groups, whereas the n-3/n-6 ratio is abnormal low after Kasai operation despite of normal function. This seems to be one factor of persisting inflammation in the liver.

Rise in contents of saturated fatty acids and lowering of the proportion of unsaturated fatty acids in blood was shown to be a universal aspect of lipid metabolism disturbance in cholelithiasis patients in genetically different populations (Tsukanov et al., 2008).

In experiments with high cholesterol fed rats, dietary administration of DHA, unlike that of eicosapentaenoic acid (EPA) 20:5n-3, significantly increase platelet membrane fluidity, evaluate by fluorescence polarization (Hashimoto et al., 2006). The comparison of the effect of DHA and EPA on platelet membrane fluidity showed that the unsaturated index and the ratio of saturated to PUFA of the platelet membrane increase only in DHA-fed rats. The authors suggest that DHA is a more potent membrane-fluidizer than EPA in platelet membrane fluidity and stronger ameliorative modulator of platelet hyperaggregation.

Effect of EPA and DHA on plasma membrane fluidity of endothelial cells cultured from the thoracic aorta by determining fluorescence polarization of 1,6-diphenyl-1,3,5-hexatriene was reported (Hashimoto et al., 1999). Plasma membrane fluidity assesses by DPH polarization is significantly higher in the order of DHA > EPA. Total phospholipid content remains unaltered in the plasma membranes of the three groups of cells; however, the molar ratio of total cholesterol to phospholipid decreases significantly only in the membranes of DHA-treated endothelial cells. The unsaturation index in the plasma membranes of EPA- and DHA-treated cells increased by 35.7 and 64.3%, respectively, compared with that in the plasma membranes of control cells. These results indicate that the influence of DHA is greater than that of EPA in increasing plasma membrane fluidity of vascular endothelial cells. The authors speculate that the greater effect of DHA compared to EPA is due to its greater ability to decrease membrane cholesterol content or the cholesterol to phospholipid molar ratio, or both, and also to its greater ability in elevating the UI in the plasma membranes of endothelial cells.

Table 12. Frequency of abnormal serum lipid values (% of the total subjects) in the patients with cardiovascular and hepatobiliary diseases with respect to normal values in healthy subjects

Serum lipids, mmol/L^{-1}	Deviation from the norm	Men				Women			
		CVD		HBD		CVD		HBD	
		%	Mean ± SE	%	Mean ± SE	%	Mean ± SE	%	Mean ± SE
Total cholesterol	–	13	4.0±0.14	35	3.7±0.11	15	3.8±0.15	30	3.8±0.06
	+	78	6.3±0.23	58	6.0±0.16	79	6.0±0.24	67	6.0±0.09
TAG	–	31	0.8±0.08	44	0.7±0.05	47	0.8±0.06	64	0.7±0.02
	+	69	2.2±0.11	44	2.1±0.14	47	2.0±0.21	29	2.0±0.14
HDL-C	–	81	0.8±0.05	41	0.9±0.04	37	0.7±0.1	39	0.9±0.03
	+	19	1.9±0.21	57	1.6±0.04	63	2.0±0.11	54	2.0±0.05
VLDL-C	–	31	0.4±0.04	44	0.3±0.02	47	0.4±0.03	63	0.3±0.01
	+	69	1.0±0.05	44	1.0±0.06	47	1.0±0.10	29	1.0±0.06
LDL-C	–	21	2.4±0.08	31	1.7±0.14	38	1.8±0.09	33	1.9±0.07
	+	76	4.2±0.18	57	3.7±0.12	63	5.0±0.32	63	4.0±0.12
Apo-B	–			35	57.3±5.86			41	56.3±2.43
	+			65	130.5±9.6			59	149.0±12.5

To expose the specific features of dislepidemia formation and development under pathology of the hepatobiliary system, the frequency of metabolic disturbances was analyzed (Tables 12, 13). About 40% of HBD patients had the normal reference level of serum lipids, the rest 60% revealed abnormal values for various lipid fractions. With that, the HBD and CVD patient groups had a similar pattern and frequency of lipid disorders. Increased values of total cholesterol, LDL cholesterol, apo B were of particularly frequent occurrence at HB pathologies. The same regularity was revealed also for CVD patients. In a majority of HBD patients, hypercholesterolemia (total cholesterol values from 5.2 to 6.5 mmol/L) was associated with an increased LDL cholesterol content and accumulation of apo B on the background of the normal level of TAG, VLDL cholesterol, and HDL cholesterol. Increased total cholesterol content in the blood of patients indicates disturbance of PUFA transport in LDL cholesterol esters, that results in their deficit in cells. At the same time, LDL accumulated in blood. This is confirmed by a distinct increase of apo B content in patients' blood; the rise is caused by accumulation of unabsorbed LDL. This effect is an evidence of obstructed PUFA transport into blood cells, a sequence of inhibition of apoB 100 receptor endocytosis.

Liver disease is accompanied by major quantitative and qualitative modifications in plasma lipoprotein metabolism. Alterations in plasma lipoprotein composition and a lower susceptibility to *in vitro* peroxidation of LDL and erythrocyte membranes were observed in liver cirrhosis (Taus et al., 1999). A rare occurrence of high TAG values is characteristic of HBD patients. Hypertriglyceridemia occurred at HBD 1.6 times less frequently that at CVD. In general, disturbance in lipid metabolism connected with increased values of total cholesterol, TAG, and apo B on the background of low contents of HDL cholesterol were in HBD patients less frequent than in CVD patients.

In the group with HBD compared with the control group, we found the decrease in PC that predominates in the outer monolayer of membranes and increase in PE and PS that are chiefly limited to the inner monolayer in normal red cells. (Table 13). Distribution of phospholipids in membranes is related to each lipid regulating various biological events through interaction with other molecules (Yamaji-Hasegama, Tsujimoto, 2006). The depletion of erythrocyte membranes in PC, which is mainly located on the outer membrane lipid layer of the cell lipid matrix, is the indication of destructive changes in the cell membrane. A consequent change of physical-chemical properties of erythrocyte membranes may lead to the emergence of new conductance channels and accelerated transport of calcium ions. This stimulates phospolipases and several other enzymes of cellular metabolism.

We found low content of PC in 90% male HBD patients examined, this is significantly more frequent against 52% in CVD group. The PC depletion in 81% HBD patients was accompanied by a rise of PS content in erythrocyte membranes. In the group of patients with CVD, changes in the PS content had an opposite direction, low PS levels dominated in 63% of patients.

In the fatty acid content of erythrocyte membranes of male patients, high levels of the fatty acids of the n-6 family (18:2n-6, 18:3n-6, and 22:5n-6) found in

100% of the examined patients with HBD (the situation does not occur in CVD patients) can be recognized as specific indications of HB pathology. The n-3 essential fatty acid depletion was reflected by increased levels of n-6 PUFA. It results in the decrease of total unsaturation index (UI).

Table 13. Composition of phospholipids and fatty acids of the erythrocyte membranes in patients with CVD and HBD. Frequency of deviation from the norm (– reduced, + elevated level).

Lipids	Vector	Men				Women			
		CVD		HBD		CVD		HBD	
		%	Mean ± SE	%	Mean ± SE	%	Mean ± SE	%	Mean ± SE
	–	63	8.50±0.47	18	9.7±0.1	57	7.8±0.51	12	7.53±0.13
PS	+	31	12.08±0.56	81	12.55±0.39	42	11.36±0.66	88	13.38±0.35
	–	63	22.12±1.18	36	24.22±0.27	100	22.75±1.30	96	24.62±0.63
SM	+	31	29.25±0.47	54	28.25±0.31	0		4	33.6
	–	52	31.9±0.61	90	30.88±0.45	71	32.42±0.76	84	32.20±0.44
PC	+	42	38.9±1.46	9	35.0±0.03	28	36.85±0.50	16	38.55±1.3
	–	15	27.2±0.20	9	23.8	0		0	
PE	+	84	31.29±0.69	90	30.75±0.32	100	34.14±1.68	100	29.14±0.61
	–	73	1.08±0.02	90	1.05±0.02	100	1.01±0.04	100	1.15±0.04
PC/PE	+	26	1.31±0.04	9	1.47±0.13	0		0	
	–	0		0		0		32	20.7±0.17
16:0	+	10	23.28±0.49	100	21.96±0.47	93	24.01±0.51	64	23.18±0.61
	–	42	0.8±0.08	40	1.1±0.11	46	0.94±0.02	4	0.4±0.01
16:1n-9	+	28	1.71±0.16	30	1.9±0.05	53	1.37±0.13	96	1.71±0.14
	–	23	14.5±0.21	20	14.5±0.3	46	15±0.48	56	14.62±0.42
18:0	+	76	17.27±0.25	80	17.16±0.51	40	17.28±0.15	36	17.67±0.27
	–	14	13.5±0.1	20	13.3±0.3	26	13.37±0.33	24	13.7±0.17
18:1n-9	+	80	15.7±0.28	80	15.58±0.63	66	15.73±0.39	68	16.15±0.35
	–	19	10.15±0.19	0		60	11.88±0.28	20	11.44±0.33
18:2n-6	+	52	13.03±0.27	100	13.62±0.42	40	14.28±0.57	64	16.25±0.88
	–	61	0.05±0.002	0		60	0.07±0.01	25	0.2±0.01
18:3n-6	+	38	0.32±0.01	100	0.3±0.02	13	0.4±0.01	12.5	0.6±0.03
	–	28	9.38±0.31	30	12.81±0.35	67	10.2±0.4	56	9.8±0.5
20:4n-6	+	61	12.32±0.24	70	9.7±0.17	33	13.0±0.1	40	13.0±0.1
	–	57	0.92±0.08	30	1.76±0.03	20	0.6±0.3	29	0.7
20:5n-3	+	23	1.88±0.23	30	0.63±0.23	73	2.0±0.23	50	1.0±0.13
	–	28	1.4±0.1	40	2.27±0.08	67	1.6±0.1	56	1.4±0.1
22:4n-6	+	52	2.29±0.13	40	1.47±0.06	20	2.0±0.23	28	2.0±0.23
	–	57	1.99±0.09	70	2.73±0.17	53	1.9±0.1	60	1.7±0.1
22:5n-3	+	28	2.56±0.09	30	2.02±0.05	40	3.0±0.23	32	3.0±0.23
	–	52	0.05±0.006	0		93	0.10±0.01	0	
22:5n-6	+	19	0.3±0.03	100	1.4±0.23	6	0.3±0.03	100	0.37±0.04
	–		5.58±0.28	100	5.74±0.31	100	5.72±0.40	100	5.21±0.22
22:6n-3	+	100		0		0		0	
	–		9.42±0.35	100	9.51±0.45	86	9.32±0.63	96	8.48±0.33
Σn-3	+	100		0		13	12.45±0.35	4	13.0±0.23
	–	9	22.1±1.9	0		80	27.18±0.29	48	27.16±0.63
Σn-6	+	80	28.42±0.45	90	30.41±0.74	13	32.32±0.37	48	33.35±0.51
	–		0.34±0.01	100	0.32±0.01	66	0.30±0.02	92	0.27±0.01
n-3\n-6	+	100		0		33	0.43±0.006	4	0.46±0.01
	–	85	151.9±2.8	80	156.7±2.1	86	152.8±4.1	92	154.0±3.1
UI	+	9	174.9±0.4	20	179.4±3.2	13	179.8±3.4	8	177.3±1.0

The fatty acid ratios may be used as indices of enzyme activity (see section Diagnostic information value of human blood fatty acids). The decrease in the concentration of n-3 PUFA correlated with an increase in the amount of docosapentaenoic acid 22:5n-6 in erythrocyte membrane. We suggest that enhanced synthesis of 22:5n-6 is likely to reflect the desaturase activity and support the fluidity level in the cells and thus the cell membrane function. This is confirmed by an analysis of metabolic conversions of fatty acids (Table 14). Indeed, the ratio of 18:2/22:5n-6, which is showing the activity of the enzymes of n-6 fatty acid biosynthesis, was twice lower than in the healthy subjects ($p < 0.001$). In contrast, this ratio increased in the CVD patients. On the other hand, a decline of unsaturation of membrane lipids in male patients with HBD shows a PUFA deficiency in the cells and confirms the suggestion on inhibition of their active transport into blood cells.

Table 14. The rations of polyunsaturated fatty acids as indices of metabolic conversions in subjects examined, mean ± SE

Ratio	Healthy n=40	Patients with CVD (control group) n=36	Patients with HBD n=35
18:2n-6 / 22:5n-6	55.48 ± 1.00	94.54 ± 2.18***	24.87 ± 3.12***
18:3n-3 / 22:6n-3	0.02 ± 0.01	0.04 ± 0.01	0.06 ± 0.01
18:4n-3 / 18:3n-3	4.94 ± 1.18	0.55 ± 0.05***	0.78 ± 0.11***
20:4n6 / 20:3n6	10.21 ± 0.37	8.51 ± 0.34**	10.10 ± 0.39
22:6n3 / 22:5n3	3.88 ± 0.13	2.53 ± 0.06***	2.57 ± 0.07***
20:4n6 / 22:6n3	1.30 ± 0.04	2.11 ± 0.09***	2.17 ± 0.06***
(20:3+20:5) / 22:6n3	0.27 ± 0.01	0.51 ± 0.01***	0.42 ± 0.01***
20:4n-6 / 20:5n-3	11.27 ± 0.70	23.11 ± 9.80	12.27 ± 1.09
20:4n-6 / 22:4n-6	6.39 ± 0.26	6.22 ± 0.24	6.76 ± 0.28
20:5n-3 / 22:5n-3	0.51 ± 0.04	0.611 ± 0.07	0.53 ± 0.03

Note: Asterisks indicate significant differences of ratios between the patients with HBD and control healthy subjects: ** for $p < 0.01$; *** for $p < 0.001$

In the female group, the phospholipid composition of the erythrocyte membranes was distinguished by low values of SM that are of importance in regulation of blood coagulation processes. This disorder was revealed in 96% of women with HBD. A decline of the SM content was associated with changes in the relative content of PC in 84% and with an increase of PS and PE in 88-100% of female patients. PS is involved in blood coagulation, apoptosis and other biological phenomena. The similar changes in the phospholipid distribution in erythrocyte membranes and the frequency of disorders were found in the women with CVD. However, high levels of PC occurred in female patients more rarely at CVD (only 42% of cases) than at HBD. The PC content is responsible for adaptive-compensatory effects on the structure and function of biomembranes, the effects are exerted through stimulation of electrokinetic and inhibition of thromboplastic properties of cells, aggregation and agglutination of erythrocytes. No doubt that increased values of 16:1 and 22:5n-6 can be assigned to as specific disturbances of the fatty acid composition in women with HBD, they are revealed in 96-100% of cases. In the CVD group, on the contrary, low values of these fatty acids were found in every female patient (100% of the examined subjects).

In summary, we conclude that the indicators of structural and functional changes at HBD are: high concentrations of PE (above 30.4%) and PS (above 12%) at low PC values, as well as an increased content of FAs of the n-6 series, such as 18:2n-6 (above 13.4%) and 22:5n-6 (above 1.4%), along with a low ratio 18:2n-6 to 22:5n-6 (below 25%) for men; and an additional evaluation criterion for female patients is a decline of the SM content (below 24.0%), along with increased contents of 16:1 up to 1.85% and of 22:5n-6 up to 0.4% and more.

Thus, it was found that the pathology of the hepatobiliary system was associated with increased levels of total cholesterol and LDL cholesterol, as well as with suppression (inhibition) of apo B-100 receptor endocytosis, which lead to block of PUFA intake into the cells. These alterations were associated with some changes in the phospholipid composition of the erythrocyte membrane: decrease in choline-containing phospholipids (PC and SM) and increase in the PS and PE contents. In the group with HBD compared with the control group, we found the decrease in PC that predominates in the outer monolayer and increase in PE and PS, which are chiefly limited to the inner monolayer in normal red cells. (Table 10). Distribution of phospholipids in membranes is related to each lipid regulating various biological events through interaction with other molecules. In eukaryotic cells, the biological membrane is characterized by a non-uniform distribution of membrane lipids, vertically as well as laterally. Such asymmetrical distribution is related to each lipid regulating various biological events through interaction with other molecules. Aminophospholipids, including PS and PE, are generally enriched in the cytoplasmic leaflet (Yamaji-Hasegama, Tsujimoto, 2006). PS is known to be involved in apoptosis, blood coagulation and other biological phenomena. The exposure of PS at the cell surface plays a critical role in blood coagulation and serves as a macrophage recognition moiety for the engulfment of apoptotic cells (Wolfs et al., 2009). Exposure of new lipid-head groups on a membrane leaflet can serve as a specific signal and, upon recognition, can be case of physiological modifications. Membrane bending is one of the mechanisms by which such activities can be triggered. The lateral membrane tension was show to be an important physical factor for the regulation of the size of the membrane invaginations (Devaux et al., 2008). We suggest that this phospholipid modification forms new structures, which are associated with diverse physiological functions in the cells such as lipid - protein interactions.

Additionally, linoleic acid underwent desaturation and chain elongation reactions, producing other essential long-chain n-6 fatty acids. The defects revealed in the phospholipid matrix of erythrocyte membranes occur also in membranes of other cells of the human organism; they are a substantial factor of disturbance of the barrier function of the cells and membrane permeability. The alteration observed in the erythrocyte membrane may result to a partial decrease or even loss of the ability of deformability, i.e. the ability to microcirculation. So, this state of the erythrocytes causes weakening of the hydrophobic bonds, impairment of physical and chemical properties of the membrane. Changes of the lipid composition of the biological membranes affect the function of integral proteins and causes destabilization of lipid-lipid and protein-lipid interactions within the membrane. Consequently, it promotes permeability of the membrane for Na^+, K^+, and Ca^{+2} ions.

A decrease in the degree of unsaturation of erythrocyte membrane lipids could thus be expected to accompany a decrease of both membrane fluidity and the activity of its membrane-bound enzymes. Since the erythrocyte membrane serves as a variable barrier to oxygen transport, the changes in its composition can induce cellular hypoxia in the tissue bed. Further, the size, shape, and diffusion capacity of a red blood cell depend on the structure of its membrane; alterations in membrane structure could therefore lead to a decrease in tissue oxygenation (Buchwald et al., 2000). These findings suggest an impairment of erythrocyte membrane physical-chemical properties in HBD as a consequence of alteration of membrane phospholipids and fatty acids. Such changes of red cell membrane might be a part of a pathogenetic mechanism responsible for hepatobiliary disorders.

Liver damage and alterations in the exocrine function of the gland lead to a profound alteration of the plasma lipoprotein profile. Abnormal hepatobiliary and circulating lipid metabolism in the Long-Evans Cinnamon (LEC) rat model of the Wilson's disease were studied (Levy et al., 2007). LEC rats displayed increased content of TAG, free sterol and cholesteryl ester in the liver. Hepatic concentrations of malondialdehyde, an index of lipid peroxidation, were also significantly elevated. This steatosis was associated with aberrant microsomal apo B-100 content, hypotriglyceridemia, hypocholesterolemia and abnormalities in both circulating lipoprotein composition and size.

Hepatic disease results in changes in the lipoprotein fatty acid composition (Arranz et al., 1996). The profound changes in the lipoprotein profile and the lower abundance of PUFA were shown in patients with hepatobiliary pathology (shuch as liver cirrhosis, uncompensated lover cirrhosis, primary biliary cirrhosis, other intrahepatic cholestasis, and acute viral hepatitis)

A decrease in the degree of polyunsaturation of lipids of erythrocyte membrane could thus be expected together with a decrease of both membrane fluidity and the activity of its membrane-bound enzymes. Because the erythrocyte membrane serves as a variable barrier to oxygen transport, the changes in its composition can induce cellular hypoxia in the tissue bed. Furthermore, because the size, shape, and diffusion capacity of a red blood cell depend on the structure of its membrane, alterations in membrane structure could lead to a decrease in tissue oxygenation (Buchwald et al., 2000). Such modification of red cell membrane might be part of part of a pathologenetic mechanism responsible for hepatobiliary pathology.

It is proposed that the broad health benefits of very long chain n-3 PUFA and the pleiotropic effects of dietary fish oil and DHA involve alterations in membrane structure and function, eicosanoid metabolism, gene expression and the formation of lipid peroxidation products. The understanding of the mechanisms of action of lipid and fatty acids by which they reduce risk and inflammatory disorders was offer by Chapkin et al. (2008). The authors present data demonstrating that DHA selectively modulates the subcellular localization of lipidated signaling proteins depending on their transport pathway, which may be universally applied to other lipidated protein trafficking. An interesting possibility raised by the current observations is that lipidated proteins may exhibit different subcellular distribution profiles in various tissues, which contain a distinct membrane lipid composition. In addition, the current findings clearly indicate that subcellular localization of proteins with a certain trafficking pathway can be subjected to selective regulation by dietary manipulation. This form of regulated plasma membrane targeting of a select subset of upstream signaling proteins may provide cells with the flexibility to coordinate the arrangement of signaling translators on the cell surface. Ultimately, this may allow organ systems such as the colon to optimally decode, respond, and adapt to the vagaries of an ever-changing extracellular environment.

In conclusion, a growing body of epidemiological, clinical, and experimental evidence has underscored the biochemical potential and diagnostic value of the fatty acid composition. We determine normal values for plasma lipid and erythrocyte phospholipid and fatty acid composition and changes that can be expected in response to hepatobiliary diseases. Serum biochemical parameters and erythrocyte membrane lipids constitute an important diagnostic tool to assess the health status, but to interpret the outcome properly. The analysis of the lipid composition of plasma and erythrocyte membranes could provide a useful tool to monitor the evolution of the hepatobiliary diseases as well as the other therapeutic pathologies.

References

Abdel-Ghaffar, TY; Amin, M; Rasheed, MA; Fouad, HH. Essential fatty acids status in infants and children with chronic liver disease. *Journal Of Clinical Biochemistry And Nutrition*, 2003 33, 1-6.

Adams, PB; Lawson, S; Sanigorski, A; Sinclair, AJ. Arachidonic acid to eicosapentaenoic acid ratio in blood correlates positively with clinical symptoms of depression. *Lipids*, 1996 31, 157-161.

Al, MD; Badart-Smook A; von Houwelingen, AC et al. Fat intake of women during normal pregnancy: relationship with maternal and neonatal essential fatty acid status. J. Am. Coll. Nutr. 1996 15, 49-55.

Albert, CM; Campos, H; Stampfer, MJ. Blood levels of long-chain n–3 fatty acids and the risk of sudden death. *N Engl J Med* 2002 346, 1113–1118.

Anderson, BM; Ma, DWL. Are all n-3 polyunsaturated fatty acids created equal? *Lipids In Health And Disease*, 2009 8, 33

Andoh, T; Yoshimitsu, M; Hatano, H. et al. The se differences in coed-blood cholesterol and fatty acid levels among Japanese fetuses.J. Epidemiol. 1997 7, 226-231.

Antoku, Y; Tsukamoto, K; Miyoshi, Y; Nagino, H; Anezaki, M; Suwa, K; Narabe, Y. Correlations of elevated levels of hexacosanoate in erythrocyte membranes with risk factors for atherosclerosis. *Atherosclerosis*, 2000 153, 169-173.

Arranz, MI; Lasuncion, MA;, Perales, J; Herrera, E; Lorenzo, I; Carcamo, C; Concostrina, L; Villar, J; Gasalla, R. Fatty acid composition of lipoprotein lipids in hepatobiliary diseases. *European Journal Of Clinical Chemistry And Clinical Biochemistry*, 1996 34, 701-709.

Bagga, D; Capone, S; Wang, HJ; Heber, D; Lill, M; Chap, L; Glaspy, JA. Dietary modulation of omega-3/omega-6 polyunsaturated fatty acid ratios in patients with breast cancer. *J. Natl. Canc. Inst.* 1997 89, 1123–1131.

Bjerve, KS; Fougner, KJ; Midthjell, K; Bobao, K. N-3 fatty acids in old age. *J. Intern. Med. Suppl.* 1989 225, 191-196.

Block, RC; Harris, WS; Reid, KJ; Sands, SA; Spertus, JA. EPA and DHA in blood cell membranes from acute coronary syndrome patients and controls. *Atherosclerosis* 2008 197, 821-828.

BoltonSmith, C; Woodward, M; Tavendale, R. Evidence for age-related differences in the fatty acid composition of human adipose tissue, independent of diet. *European Journal Of Clinical Nutrition*, 1997 51, 619-624.

Boulton, TJC; Magarey, AM; Cockington, RA. Serum lipids and apolipoprotiens from 1 to 15 years: changes with age and puberty, and relationships with diet, parental cholesterol and family history of ischaemic heart disease. Acta Pediatrica. 1995 84, 1113-1118.

Buchwald, H; Menchaca, HJ;. Michalek, VN; Rohde, TD; Buchwald, DB; Hunninghake, HJ; O'Dea TJ. Plasma cholesterol: an influencing factor in red blood cell oxygen release and cellular oxygen availability. *J. Am. Coll. Surg.* 2000 191, 490–497.

Caggiula, AW; Mustad, VA. Effects of dietary fat and fatty acids on coronary artery disease risk and total and lipoprotein cholesterol concentrations: epidemiologic studies. *Am. J. Clin. Nutr.* 1997 65, 1597S–1610S.

Calder, PC. Polyunsaturated fatty acids, inflammation, and immunity. *Lipids* 2001 36, 1007–1024.

Chapkin, RS; Seo, JM; McMurray; DN, Lupton, JR. Mechanisms by which docosahexaenoic acid and related fatty acids reduce colon cancer risk and inflammatory disorders of the intestine. *Chemistry And Physics Of Lipids*, 2008 153, 14-23.

Cooper, RA. Lipids of human red cell membrane: normal composition and variability in disease. *Semin. Hematol.*, 1970 7, 296–322.

Das, UN. Essential fatty acid metabolism in patients with essential hypertension, diabetes mellitus and coronary heart disease. *Prostaglandins, Leukotrienes and Essential Fatty Acids,* 1995 53, 387-391.

Decsi, T; Kolechko, B. Fatty-acid composition of plasma-lipid classes in healthy-subjects from birth to young adulthood. *European Journal Of Pediatrics*, 1994 153, 520-525.

Decsi, T; Kolechko, B. Role of long-chain polyunsaturated fatty acids in early human neurodevelopment. *Nutr. Neurosc.* 2000 3, 293–306.

Devaux, PF; Herrmann, A; Ohlwein; N, Koziov, MM. How lipid flippases can modulate membrane structure. *Biochimica Et Biophysica Acta-Biomembranes*, 2008 1778, 1591-1600.

Dewailly, E; Blanchet, C; Lemieux, S. N–3 fatty acids and cardiovascular disease risk factors among the Inuit of Nunavik. *Am J Clin Nutr* 2001 74, 464–73.

Endakova, EA; Novgorodtseva, TP; Svetashev, VI. Modification of blood fatty acid composition in case of cardio-vascular diseases 2002 Dalnauka. Vladivostok. 295 P.

Fekete, K; Marosvolgyi, T; Jakobik, V; Decsi, T. Methods of assessment of n-3 long-chain polyunsaturated fatty acid status in humans: a systematic review. *American Journal Of Clinical Nutrition*, 2009 89, 2070S-2084S.

Fenton, WS; Hibbeln, J; Knable, M. Essential fatty acids, lipid membrane abnormalities, and the diagnosis and treatment of schizophrenia. *Biological Psychiatry*, 2000 47, 8-21.

Galland, L. Increased requirements for essential fatty acids in atopic individuals: a review with clinical descriptions. J. *Am. Coll. Nutr*. 1986 5, 213-228.

Gerique, JAG; Porres, A; Martinez, DL; et al. Levels of lipoprotein and plasma lipids in Spanish children aged from 4 to 18 years. *Acta Pediatrica*. 1996 85, 38-42.

Gibson, RA; Sinclair, AJ. Are Eskimos obligate carnivores? *Lancet*. 1981 1, 1100.

Guerra, A; Demmelmair, H; Toschke, AM; Koletzko. B. Three-year tracking of fatty acid composition of plasma phospholipids in healthy children. *Annals Of Nutrition And Metabolism*, 2007 51, 433-438.

Harris, WS. The omega-3 index as a risk factor for coronary heart disease *Am. J. Clinical Nutrition* 2008 87, 1997S-2002S.

Harris, WS; Sands, SA; Windsor, SL; et al. Omega-3 fatty acids in cardiac biopsies from heart transplant patients: correlation with erythrocytes and response to supplementation. *Circulation* 2004 110, 1645-1649.

Harris, WS; von Schacky, C. The Omega-3 Index: a new risk factor for death from coronary heart disease? *Prev Med* 2004 39, 212–220.

Harwood, J.L. Lipid metabolism The Lipid Handbook Eds. Gunstone, FD; Harwood JL; Padzey, FB. London, Glasgow, New York, Tokyo, Melbourne, Madras, 1994 605-633.

Hashimoto, M; Hossain, S; Shido, O. Docosahexaenoic acid but not eicosapentaenoic acid withstands dietary cholesterol-induced decreases in platelet membrane fluidity. *Molecular And Cellular Biochemistry,* 2006 293, 1-8.

Hegele, RS. Genetic variation of intestinal fatty acid – binding protein associatioed with variation in body mass in aboriginal Canadians. *J. Clin. Endocrinol. Metab*. 1996 81, 4334-4337.

Hoffman, DR; Birch, EE; Birch DG; et al. Effect of supplementation with ω-3 long-chain polyunsaturated fatty acids on retinal and cortical development in premature infants. *Am. J. Clin. Nutr*. 1993 57, 807S-812S.

Horrobin, DF. Low prevalences of coronary heart disease (CHD), psoriasis, astma and rheumatoid arthritis in Eskimos: are they cause by high dietaryintake of eicosapentaenoic acid (EPA), a genetic variation of essential fatty acid (EFA) metabolism or a combination of both? *Med. Hypoth*. 1987 22, 421-428.

Horton, TJ; Commerford, SR; Pagliassotti, MJ; Bessesen, DH. Postprandial leg uptake of triglyceride is greater in women than in men. *Am. J. Physiol. Endocrinol. Metab*. 2002 283, E1192-1202.

Innis,SM; Owens, SD. Dietary fatty acid composition in pregnancy alters neurite membrane fatty acids and dopamine in newborn rat brain. *Journal Of Nutrition* 2001 131, 118-122.

Jakobik, V; Burus, I; Decsi, T. Fatty acid composition of erythrocyte membrane lipids in healthy subjects from birth to young adulthood. *European Journal Of Pediatrics*, 2009 168, 141-147.

Jensen, MD; Cryer, PE; Johnson, CM; Murray, MJ. Effects of epinephrine on regional free fatty acid and energy metabolism in men and women. *Am. J. Phisiol*. 1996 270, E259-264.

Kabagambe, EK; Tsai, MY; Hopkins, PN; Ordovas, JM; Peacock, JM; Boreck, IB; Arnett, DK. Erythrocyte fatty acid composition and the metabolic syndrome: a national heart, lung, and blood institute GOLDN study. *Clinical Chemistry* 2008 54, 154–162.

Karpe, F. Mechanisms of postprandial hyperlididemia-remnants and coronary artery disease. *Diabet Med*. 1997 14, S60-66.

Klimov, AN; Nagorvev, VA. Mechanism of lipoprotein penetration into the arterial-wall leading to development of atherosclerosis. *Atherosclerosis Reviews* 1983 11, 107-156.

Knuth, ND; Horowitz, JF. The elevation of ingested lipids within plasma chylomicrons is prolonged in men compared with women. *J Nutr*, 2006 136, 1498-1503.

Krachler, B; Norberg, M; Eriksson, JW; Hallmans, G; Johansson, I; Vessby, B; Weinehall, L; Lindahl, B. Fatty acid profile of the erythrocyte membrane preceding development of Type 2 diabetes mellitus. *Nutrition Metabolism And Cardiovascular Diseases*, 2008 18, 503-510.

Krause, C; Rosewich, H; Gartner, J. Rational diagnostic strategy for Zellweger syndrome spectrum patients. *European Journal Of Human Genetics*, 2009 17, 741-748.

Lambert, J; Agostoni ,C; Elmadfa, I; Hulshof, K; Krause, E;, Livingstone, B; Socha, P; Pannemans, D; Samartin, S. Dietary intake and nutritional status of children and adolescents in Europe. *British Journal Of Nutrition*, 2004 92, S147-S211.

Lausada, NR; Boullon, S; Boullon, F; De Gomez Dumm, IN. Erythrocyte membrane, plasma and atherosclerotic plaque lipid pattern in coronary heart disease. *Medicina-Buenos Aires*, 2007 67, 451-457.

Lebedeva, EN. Dislipidemia and hypovitaminosis in children of school age. PhD thesis. 1994.

Lemaitre, RN; King, IB; Raghunathan, TE; et al. Cell membrane trans-fatty acids and the risk of primary cardiac arrest. *Circulation* 2002 105, 697–701.

Levant, B; Ozias, MK; Carlson, SE. Diet (n-3) polyunsaturated fatty acid content and parity affect liver and erythrocyte phospholipid fatty acid composition in female rats. *J. Nutr.* 2007 137, 2425-2430.

Levy, E; Brunet, S; Alvarez, F; Seidman, E; Bouchard, G; Escobar, E; Martin, S. Abnormal hepatobiliary and circulating lipid metabolism in the Long-Evans Cinnamon rat model of Wilson's disease. *Life Sciences* 2007 80, 1472-1483.

Magkos, F ; Mittendorfer, B. Gender differences in lipid metabolism and the effect of obesity. *Obstetrics And Gynecology Clinics Of North America*, 2009, 36, 245-254.

Malinska, H; Oliyarnyk, O; Hubova, M; Zidek, V; Landa, V; Simakova, M; Mlejnek, P; Kazdova, L; Kurtz, TW; Pravenec, M. Increased liver oxidative stress and altered PUFA metabolism precede development of non-alcoholic steatohepatitis in SREBP-1a transgenic spontaneously hypertensive rats with genetic predisposition to hepatic steatosis. *Mol Cell. Biochem.* 2009 In press.

Minda, H; Larque, E; Koletzko, B; Decsi, T. Systematic review of fatty acid composition of plasma phospholipids of venous cord blood in full-term infants. *European Journal Of Nutrition*, 2002 41, 125-131.

Montagne, CT; O'Rahilly, S. The perils of portliness - Causes and consequences of visceral adiposity. *Diabetes* 2000 49, 883-888.

Nagao, K; Yanagita, T. Conjugated fatty acids in food and their health benefits. *J. Biosci. Bioeng.* 2005 100, 152–157.

Nettleton, J.A. Omega-3 fatty acids and health. New York. Chapman and Hall. 1995. P. 361.

Nijs, J; De Meirleir, K. Oxidative stress might reduce essential fatty acids in erythrocyte membranes of chronic fatigue syndrome patients. *Nutritional Neuroscience*, 2004 7, 251-253.

Novgorodtseva, TP. Erythrocyte phospholipids at development of coronary heart disease. *The Bull of Siberian Branch of RAMS*. 1998 1, 15-18.

Novgorodtseva, TP; Abakumov, AI; Sorokina, LV. Multidimensional statistic metods and the diagnostic significance of erythrocyte fatty acids in cardiovascular diseases. *Klin. Lab. Diag*. 2001. 3:13-16.

Novgorodtseva, TP; Ivanov, EM; Antoniuk, MV; Kamanan, YK; Zhukova , NV; Yurenko, AV. Non-esterified fatty acid composition in patients with metabolic syndrome. *Klin. Lab. Diag.* 2008 10, 38-40.

Novgorodtseva, TP; Kamanan, YK; Antoniuk, MV; Zhukova, NV. Modification of free and esterified fatty acids under metabolic syndrome. *Klinical Medicine* 2009 5, 33-37.

Novgorodtseva, TP; Svetashev, VI; Zhukova, NV. Ontogenetic regularities of phospholipid pool formation of rat erythrocyte membranes at normal and heart pathology development. Proceedings of Institute of Medical Climatology and Rehabilitative Treatment. Vladivostok. Dalnauka. 2004 78-87.

Novgorodtseva, TP; Viazova, AV; Zhukova, VV. Membrane lipids of patients with chronic bronchitis associated with urolithiasis in phase of clinical remission. *Klin. Lab. Diag.* 2006 2, 7-9.

Novgorodtseva, TP; Zhuravskaya, NS; Vostrikova, O.G. Blood lipids in offsprihgs of patients with ischemic heart disease. *Kardiologiia* 2002 6, 32-37.

Novgorodtseva, TP; Zhuravskaya, NS; Vostrikova, OG. Blood lipids in offsprihgs of patients with ischemic heart disease. *Kardiologiya*, 2002 42, 32-37.

Pala, V; Krogh, V; Muti, P; Chajes, V; Riboli, E; Micheli, A; Saadatian, M; Sieri,S; Berrino F. Erythrocyte membrane fatty acids and subsequent breast cancer: a prospective Italian study. J. *Natl. Cancer Inst.* 2001 93, 1088–95.

Peterson, DB; Fisher, K; Carter RD; Mann, N. Fatty acid composition of erythrocytes and plasma triglyceride and cardiovascular risk in Asian diabetic patients. *Lancet*. 1994 342, 1528-1530.

Pettei, MJ; Daftary, S; Levine, JJ. Essential fatty-acid deficiency associated with the use of a medium-chain-triglyceride infant formula in pediatric hepatobiliary disease. *American Journal Of Clinical Nutrition*, 1991 53, 1217-1221.

Poppitt S.D., Kilmartin P., Butler P., Keogh G.F. Assessment of erythrocyte phospholipid fatty acid composition as a biomarker for dietary MUFA, PUFA or saturated fatty acid intake in a controlled cross-over intervention trial. *Lipids Health Dis*. 2005 4. P. 30-37.

Rajeshwari, R; Nicklas, TA; Pownall, HJ; Berenson, GS. Cardiovascular diseases - a major health risk in Asian Indians. *Nutrition Research* 2005 25, 515-533.

Ricci, S; Patoia, L, Berretini, M. et al. Fatty acid pattern of red blood cell membranes and risk of ischemic brain infarction – a case-control study. *Stroke*. 1987 18, 575-578.

Robberecht, E; Koletzko, B; Christophe, A. Several mechanisms contribute to the abnormal fatty acid composition of serum phospholipids and cholesterol esters in cholestatic children with extrahepatic biliary atresia. *Prost Leuk Ess FA* 1997 56, 199-204.

Rosseneu, M. Cambien, F; Vinaimont, N; Nicaud, V; Debacker, G. Biomarkers of dietary-fat composition in young-adults with a parental history of premature coronary heart-disease compared with controls - the ears study.3E Atherosclerosis 1994 108, 127-136.

Salem, N.Jr.; Niebylski, CD. The nervous system has an absolute molecular species requirement for proper function. *Mol. Membr. Biol.* 1995 12, 131-134.

Sands, SA; Reid, KJ; Windsor, SL; Harris, WS. The impact of age, body mass index, and fish intake on the EPA and DHA content of human erythrocytes. *Lipids* 2005 40, 343-347

Shamir, R; Tershakovec, AM; Gallagher, PR;, Liacouras, CA; Hayman, LL; Cortner, JA. The influence of age and relative weight on the presentation of familial combined hyperlipidemia in childhood. *Atherosclerosis*, 1996 121, 85-91.

Shestopalov, EY; Novgorodtseva, TP; Kamanan, YK; Antoniuk, MV; Zhukova, NV. Status of lipid components of the erythrocyte membranes in patients with comorbidities of viscera. *Bulletin of Siberian Branch of Russian Academy of Medical Sciences* 2009 135, 32-37.

Simopoulos, AP. The importance of the omega-6/omega-3 fatty acid ratio in cardiovascular disease and other chronic diseases *Exp Biol Med* 2008 233,674-88.

Socha, P; Koletzko B, Pawlowska, J; Socha, J. Essential fatty acid status in children with cholestasis, in relation to serum bilirubin concentration. *Journal Of Pediatrics*, 1997a 131, 700-706.

Socha, P; Koletzko, B; Jankowaka, I; Pawlowska, J; Demmelmair, H; Swiatkowska, E; Socha, J. Long-chain PUFA supplementation improves PUFA profile in infants with cholestasis. *Lipids*, 2002 37, 953-957.

Socha, P; Koletzko, B; Pawlowska, J; Proszynska, K; Socha, J. Treatment of cholestatic children with water-soluble vitamin E (alpha-tocopheryl polyethylene glycol succinate): Effects on serum vitamin E, lipid peroxides, and polyunsaturated fatty acids. *Journal Of Pediatric Gastroenterology And Nutrition*, 1997b 24, 189-193.

Socha, P; Koletzko,B; Swiatkowska, E; Pawlowska, J; Stolarczyk, A; Socha, J. Essential fatty acid metabolism in infants with cholestasis. *Acta Paediatrica*, 1998 87, 278-283.

Sumida, W; Kaneko, K; Ono, Y; Tainaka, T; Ando, H. Different polyunsaturated fatty acid profiles in patients with biliary atresia after successful Kasai operation and liver transplantation. *Pediatric Surgery International*, 2009 25, 255-259.

Sumner, AE; Farmer, NM; Tulloch-Reid, MK; Sebring, NG; Yanovski , JA; Reynolds, JC; Boston, RC; Premkumar, A. Sex differences in visceral adipose tissue volume among African Americans, *Am. J. Clin. Nutr.* 2002 76, 975-979.

Sumner, AE; Kushner, H; Tulenko, TN; Falkner, B; Marsh, JB. The relationship in African-Americans of sex differences in insulin-mediated suppression of nonesterified fatty acids to sex differences in fasting triglyceride levels. *Metabolism-Clinical And Experimental*, 1997 46, 400-405.

Sun, Q; Ma, J; Campos, H; Hankinson, SE; Hu, FB. Comparison between plasma and erythrocyte fatty acid content as biomarkers of fatty acid intake in US women. *Am J Clin Nutr* 2007 86, 74–81.

Takamatsu, H; Noguchi, H; Tahara, H; Kaji, T; Shimono, R; Ikee, T; Andoh, T. Plasma fatty acids concentrations in postoperative patients with biliary atresia. *Nutrition*, 1999 15, 755-759

Taus, M; Dousset, N; Moreau, J; Ferretti, G; Galeazzi, T; Solera, ML; Curatola, G; Valdiguie, P. Study of fluidity of low density lipoproteins from liver cirrhotic patients. *Nutrition Metabolism And Cardiovascular Diseases*, 1999 9, 289-293.

Theret, N; Bard, JM; Nuttens, MC et al. The relationship between the phospholipid fatty acid composition of the red blood cells, plasma, lipids and apolipiprotiens. Metabolism. 1993 42, 562-568.

Titov, VN. Blockade of saturated fatty acids transport to a cell: Pathogenesis of essential hypertension. Voprosy Meditsinskoi Khimii. 1998 44 317-330.

Tsukanov, VV; Kuperstein, EY; Tonkikh, YL; Bronnikova, EP. Association of fatty acid composition of blood serum with bile lipid composition in cholelithiasis patients. *Terapevticheskii Arkhiv*, 2008, 80, 71-75.

Tziomalos K;, Weerasinghe, CN; Mikhailidis, DP; Seifalian, AM. Vascular risk factors in South Asians. *International Journal Of Cardiology* 2008 128, 5-16.

Uauy, R; Hoffman, DR; Mena, P; Llanos, A; Birch, EE. Term infant studies of DHA and ARA supplementation on neurodevelopment: results of randomized controlled trials. *J. Pediatr.* 2003 143, 17–25.

van Wezel-Meijler, G; van der Knaap, MS; Huisman, J; Jonkman, EJ; Valk, J; Lafeber, HN. Dietary supplementation of long-chain polyunsaturated fatty acids in preterm infants: effects on cerebral maturation. *Acta Paediatrica* 2002 91 942-950.

Vostrikova, OG; Slezka, IE. Status of lipid metabolis in children population of South of Primorye region. *Human Physiology* (in Russian) 1999 3, 132-137.

Warensjo, E; Ohrvall, M; Vessby, B. Fatty acid composition and estimated desaturase activities are associated with obesity and lifestyle variables in men and women. *Nutrition Metabolism And Cardiovascular Diseases* 2006 16, 128-136.

Williams, CM. Lipid metabolism in women. *Proceedings Of The Nutrition Society,* 2004 63, 153-160.

Wolfs, JLN; Comfurius, P; Bekers, O; Zwaal, RFA; Balasubramanian, K; Schroit; AJ, Lindhout, T; Bevers, EM. Direct inhibition of phospholipid scrambling activity in erythrocytes by potassium ions. *Cellular And Molecular Life Sciences*, 2009. 66, 314-323.

Yamaji-Hasegawa, A; Tsujimoto, M. Asymmetric distribution of phospholipids in biomembranes. *Biological & Pharmaceutical Bulletin*, 2006, 29, 1547-1553.

Yamashiro, Y; Shimizu, T; Ohtsuka, Y; Nittono, H; Miyano, T; Kawakami, S; Hayasawa, H. Docosahexaenoic acid status of patients with extrahepatic biliary atresia. *Journal Of Pediatric Surgery,* 1994 29, 1455-1458.

Zappalla, FR; Gidding, SS. Lipid Management in Children. *Endocrinology And Metabolism Clinics Of North America* 2009 38, 171-

Zeh, N; Rossi, SS; Hofmann, AF; Steinbach, JH; Hagey, LR; Oliver, D; Stange, J; Hassanein, T. The effect of extracorporeal albumin dialysis on plasma phospholipid fatty acids in patients with end-stage liver disease. *Liver International*, 2003 23, 34-38.

In: Saturated Fats ISBN: 978-1-61209-566-0
Editor: Joseph P. Langella

Chapter VI

Fatty Acids and Immune System Functions: From Biological Properties to Clinical Consequences

María A. Puertollano, Lidia Cruz-Chamorro, Elena Puertollano, María T. Pérez-Toscano, Gerardo Álvarez de Cienfuegos and Manuel A. de Pablo[*]
Unit of Microbiology, Department of Health Sciences, Faculty of Experimental Sciences, University of Jaén, Jaén, Spain

Abstract

The potential role of certain fatty acids on immune system functions has been intensively investigated during the last thirty years. Fatty acids are characterized by exerting various biological functions on the cells from mammalian. In addition to their crucial role as main source of energy or as structural substances, different types of fatty acids have been considered as

[*] Correspondence concerning this article should be addressed to Dr. Manuel Antonio de Pablo Martínez, Universidad de Jaén, Facultad de Ciencias Experimentales, Departamento de Ciencias de la Salud, Área de Microbiología E-23071-Jaén (Spain), Tel. +34 953 212 003; Fax. +34 953 212 141; e-mail address: mapablo@ujaen.es.

modulating agents of immune system in both animals and humans. Based on this preliminary argument, different mechanisms of action have been proposed in order to explain these events: i) the capacity of fatty acids to alter the membrane fluidity because they are incorporated in the plasma membrane phospholipids, ii) the formation of lipid peroxides that may be toxic to the cells, iii) the changes in the production of lipid mediators such as eicosanoids or iv) the direct intervention in gene expression. Recently, numerous studies have determined their contribution as inducers or inhibitors of apoptosis. Therefore, this last event may constitute an important explanation to clarify the reasons whereby certain fatty acids modulate the immune system. Thus, the most immunosuppressive fatty acids are derived from long-chain *n*-3 polyunsaturated fatty acids (mainly contained in fish oil) and from *n*-9 monounsaturated fatty acids (mainly contained in olive oil). These fatty acids are involved in the reduction of lymphocyte proliferation, cytokine production, natural killer (NK) cell activity, and alteration of surface molecules. The modification of these immune functions leads to a suppression of innate and adaptive immunity. Hence, these actions acquire a significant importance in the application of these fatty acids in the diminution of inflammatory disorders, which are characterized by an overactivation of immune system, or in the reduction of tumor development. In spite of the relevance of several fatty acids of the diet as beneficial agents in the human health, an excess of determined types of fatty acids may be involved in an increase of the host susceptibility to infectious microorganisms. Taking into account the consequences derived from the administration of dietary lipids, here we will review the role of dietary lipids as biological agents capable of modulating the immune system function. Moreover, we will study the mechanisms of action and the immune parameters susceptible to be modified after the incorporation of certain fatty acids in the diet. A second part of this series will analyze the clinical consequences derived from the administration of certain dietary lipids, which will be based on the intervention of fatty acids in immunonutrition; their application in clinical nutrition, their functions as modifier agents of immune resistance or their role in cancer aetiology.

Keywords: Dietary lipids, Fatty acids, Olive oil, Fish oil, Lymphocyte, Cytokines, Immunomodulation, Immune system.

1. Introduction

The relationship between nutrition and immune system has been a topic of research for more than 50 years. It is widely known that a deficiency or an excess of nutrients affect adversely the activity and the function of immune system. Among the nutrients capable of modulating the immune system functions, we can include unsaturated fatty acids, which have intensively been studied during the last years [7-9,18,19,36]. In general, fatty acids are essential components in both animals and humans; they have complex structural and functional roles in eucaryotic cells. They are mainly responsible for energetic functions or for structural functions. Hence, it is well documented the action of fatty acids as energy source, because they are excellent respiratory substrates in most types of cells. On the other hand, fatty acids represent important components in plasma cell membrane, because they participate along with other substances in the formation of membrane phospholipids and in the development of this cellular structure.

Irrespective of their main function as energetic or structural components, certain fatty acids are associated with the modulation of different immune system functions. Nevertheless, the role of fatty acids on immune system functions differs among the fatty acids. Long-chain *n*-3 polyunsaturated fatty acids (derived from fish oil, such as eicosapentaenoic acid or docosahexaenoic acid) exert the most potent immunomodulatory activities, but *n*-6 polyunsaturated fatty acids (derived from vegetables) or *n*-9 monounsaturated fatty acids (derived from olive oil, such as oleic acid) are also related to produce an immunomodulatory activity. Among the immune parameters modulated by unsaturated fatty acids are lymphocyte proliferation, cytokine secretion, natural killer (NK) cell activity or adhesion molecules. The alteration of the mentioned immune functions generally promotes a suppression of immune system.

The present review will describe the function of dietary lipids upon the modulation of immune system in both humans and animals. It examines recent findings describing the potential role of different fatty acids on immune system functions, and the mechanisms of action responsible for the modulation of immune system. Previously, we will carry out a brief description of immune system and its functionality. Subsequently, we will comment the potential action of certain fatty acids on immune system functions, as well as their mechanisms of action.

2. Immune System: A Brief Overview

Immunity is a wide concept applied generally in vertebrates, which may be defined as a state of resistance and protection against the entry of biological agents (such as viruses, bacteria, fungi or parasites) that allows the appropriate recognition of them as foreign organisms, as well as the elimination and destruction of these infectious agents. Therefore, immune system has evolved to maintain the internal environment, particularly against invading microorganisms. Moreover, immune system includes another main function, a memory component that allows a second response faster than the initial response in the case of an antigen is encountered again.

Overall, the immune system is an integrated defence network that comprises two separate but interacting and interdependent types: (i) the innate, natural or non-specific immune system or, (ii) the acquired, adaptive or specific immune system. Both components of immunity involve the production and release of different factors and the recruitment of various types of cells (leukocytes), which are originated in bone marrow. These cells circulate in the bloodstream or organised into lymphoid organs. They are divided into two broad groups: i) phagocytes divided into mononuclear cells (macrophages/monocytes) or polymorphonuclear cells (eosinophils, neutrophils and basophils), and ii) lymphocytes (T lymphocytes, B lymphocytes and natural killer cells [NK]). Similarly, T lymphocytes can be divided into two fundamental groups: on the one hand, Th1 lymphocytes, which are regulated by interleukin-12 (IL-12) and interferon-γ (IFN-γ), produce pro-inflammatory cytokines (such as IL-2 or IFN-γ) and activate macrophages, NK cells and cytotoxic T lymphocytes; on the other hand, Th2 lymphocytes which are regulated by IL-4, they are specialized in the production of cytokines with an anti-inflammatory activity (such as IL-4, IL-5 or IL-10). Hence, infection with intracellular pathogens will induce the differentiation along the Th1 pathway, whereas infection with extracellular pathogens will promote the differentiation along the Th2 pathway. Cytokines comprise the interleukins (IL), interferons (IFN), tumor necrosis factor (TNF) and transforming growth factor (TGF).

As mentioned above, the immune system is a complex network consisting of different types of cells and factors responsible for the protection of the host from infectious agents. Hence, immune system should recognise the self from non-self and to eliminate non-self (foreign agents), which constitutes the essential objective of this system, whereas the loss of this property contributes to the generation of autoimmune diseases, which are characterized by a dysregulated balance between Th1- and Th2-type response.

2.1. Innate Immunity

The innate arm of immunity prevents the entry of foreign agents into the body. It plays a crucial role in the early control of pathogenic infectious agents as well as in the initiation and subsequent course of the acquired immunity. Therefore, it constitutes the first line (early phase) of defence against pathogens. On the basis of this preliminary concept, innate immunity represents an important mechanism which rapidly eliminates infectious agents through different mechanisms such as phagocytosis or engulfment of infectious agents, direct destruction by the production of complement or toxic substances released from phagocytic cells or NK cells. In fact, innate immunity is the most efficient mechanism in order to eliminate intracellular growth pathogenic agents. Innate resistance does not distinguish among microorganisms and does not change in intensity upon re-exposure.

Table 1. Differential properties of both innate and acquired immune system

Immune system		
Factors	Innate Immunity	Acquired Immunity
Specificity	Nonspecific	Specific
Physicochemicals	Skin Acid in the stomach Sebaceous gland secretions Lysozime Natural microbiota	Antibodies in mucosa
Cells	Monocytes/macrophages Polymorphonuclear cells Natural killer cells	B lymphocytes T lymphocytes
Serum proteins	Complement	Antibodies
Cytokine production	Macrophage-derived cytokines	Lymphocyte-derived cytokines

2.2. Acquired Immunity

On the other hand, acquired immunity requires the identification of molecules from an invading agent. The recognition of antigens is carried out by B lymphocytes (which produce antibodies) and T lymphocytes which recognize the antigens on the surface of cells. B cells secrete immunoglobulins (antibodies) with identical antigen specificity to that of its surface receptor. Intracellular infections are communicated to T lymphocytes by the expression of peptide fragments along with proteins of the major histocompatibility complex (MHC). Thus, maturation and expansion of T lymphocytes is an essential process for destroying invading organisms. The link of peptide (antigen) and MHC is identified by T lymphocytes. There are two types of MHC molecules: MHC class I and MHC class II. MHC I binds intracellular peptides (from viruses or intracellular growth bacteria), and it is subsequently recognized by T lymphocytes expressing CD8 receptor (cytotoxic/suppressor T lymphocytes) which leads to the destruction of the infected cell. MHC II binds peptides from foreign agents that have been phagocytosed by macrophages or by antigen-presenting cells (extracellular antigens) and it is recognized by T lymphocytes expressing CD4 receptor (helper/inducer T lymphocytes) which leads to activating T cell-mediated response. Finally, both innate and acquired immune responses are integrated through the interactions of cells and cytokine production generated as a result of a specific stimulus. Thus, the infectious agent is eliminated and immunological memory remains; therefore, the response of the immune system against a second response by a similar antigen is more rapid, allowing the protection of host from infection [30].

Differential properties of both innate and acquired immune system are summarized in Table 1.

3. Fatty Acids and Immune System: Biological Properties

3.1. Types of fatty Acids

The main families of fatty acids are divided into two types called essential fatty acids for mammalian cells and should be supplied in the diet, because they cannot be synthesized de novo. Essential fatty acids include the *n*-3 series, which are derived from linolenic acid, and *n*-6 series, which are derived from linoleic acid. Depending on its position, the first double bond from the methyl end in unsaturated fatty acids plays an important role; thus, fatty acids may be divided into *n*-3 (when the first double bonds are between C3 and C4), *n*-6 (when the first double bonds are between C6 and C7) and *n*-9 (when the first double bonds are between C9 and C10). The three types of unsaturated fatty acids presented are not interconvertible, but they are metabolized by a common series of desaturases and elongases (Figure 1). The main *n*-3 polyunsaturated fatty acids found in dietary lipids are α-linolenic acid (18: 3*n*-3), eicosapentaenoic acid (20: 5*n*-3) or docosahexaenoic acid (22: 6*n*-3). The main *n*-6 polyunsaturated fatty acids found in dietary lipids are linoleic acid (18: 2*n*-6) and arachidonic acid, whereas other *n*-6 polyunsaturated fatty acids are less abundant than linoleic acid, such as conjugated linolenic acid and γ-linolenic acid (18: 3*n*-6). Finally, the most important *n*-9 monounsaturated fatty acid found in dietary lipids is oleic acid (18: 1*n*-9).

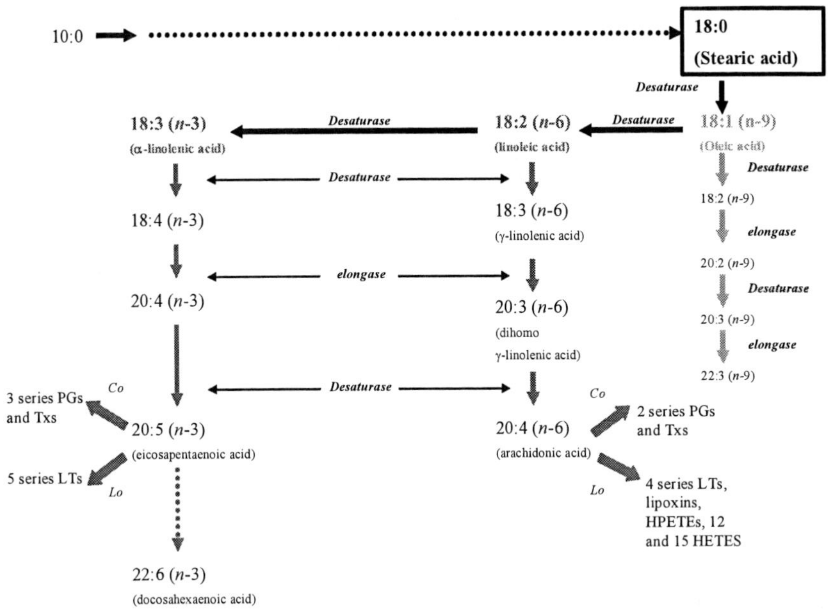

Figure 1. Different pathways of *n*-3 (blue), *n*-6 (red) or *n*-9 (green) fatty acid synthesis. Abbreviations: Co, cyclooxygenase; Lo, lipooxygenase; PGs, prostaglandins; LTs, leukotrienes; Txs, thromboxanes; hydroxy-eicosatetraenoic acids (HETEs); hydroperoxy-eicosatetraenoic acids (HPETEs).

A representation of the most important saturated and monounsaturated fatty acids is illustrated in Figure 2, and polyunsaturated fatty acids appear in Figure 3.

3.2. Mechanisms of Action

As mentioned previously, for several years numerous studies have attempted to elucidate the mechanisms whereby several dietary lipids produce a potential effect on immune system functions [43]. Although, many human studies investigating the relationship between dietary lipids and immune system have been considered to be inconsistent, several hypotheses have been suggested as possible mechanisms of action responsible for the modulation of immune system functions. Experimental investigations have confirmed that several fatty acids exert changes in the phospholipids of plasma membrane which affect the membrane fluidity [74], they also alter eicosanoid production [21], produce lipid peroxides [1], or participate in regulation of the transcription factors [35].

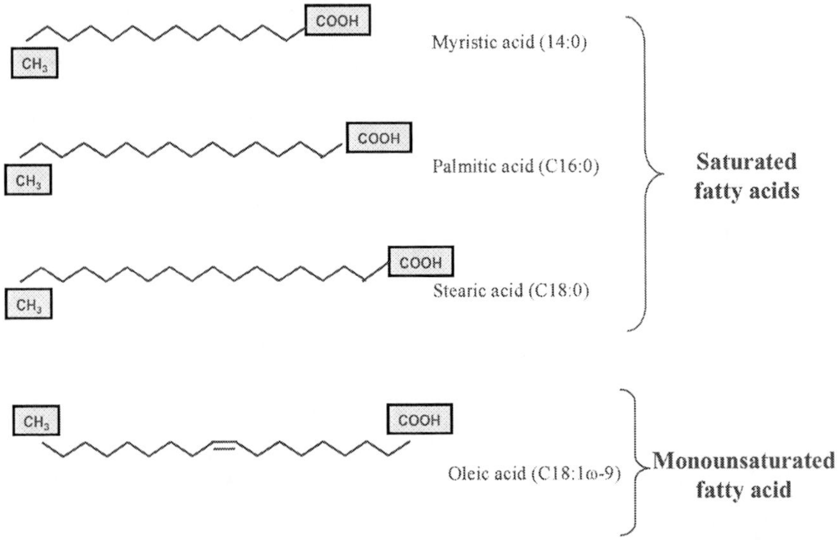

Figure 2. Representation of the most important class of saturated, and monounsaturated fatty acids. The first number indicates the number of carbon atoms in the fatty acid chain, the second number indicates the number of double bonds. The family of monounsaturated fatty acid is indicated by *n*-9.

Currently, the study of membrane lipid microdomains, so-called rafts, is acquiring a great importance, in order to explain the modulatory mechanisms of certain fatty acids on T cell signal transduction [64]. Membrane lipid rafts may be defined as specialized regions within plasma membrane composed by high concentration of cholesterol and sphingolipids (sphingomyelin and glycolipids) and a polar region that contains saturated fatty acid residues [5]. A great variety of proteins that participate in T cell signalling are concentrated in rafts and many of these proteins are palmitoylated [5]. The distribution of lipid rafts into plasma membrane differs among the cells, but the membrane surface that surround lipid rafts is composed by unsaturated fatty acids, which contribute to a relative fluidity of plasma membrane [64]. Thus, the treatment of Jurkat cells in the presence of *n*-3 polyunsaturated fatty acid, such as eicosapentaenoic acids produced a marked

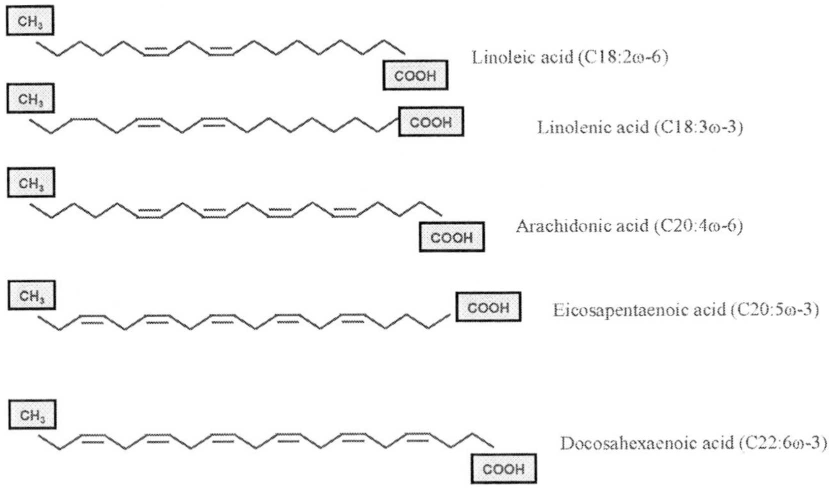

Polyunsaturated fatty acids

Figure 3. Representation of the most important class of polyunsaturated fatty acids. The first number indicates the number of carbon atoms in the fatty acid chain, the second number indicates the number of double bonds. The family of polyunsaturated fatty acid is indicated by *n*-3 or *n*-6.

enrichment of eicosapentaenoic and docosapentaenoic acid in lipids from isolated rafts [66]. Similarly, palmitate-labeled proteins and the linker for activation of T cells (LAT) were both displaced from rafts due mainly to altered raft lipid composition [66]. Numerous important proteins involved in T cell signalling are concentrated in rafts, and alterations of raft lipid fatty acyl composition is a crucial event that participates in the inhibitory effects of polyunsaturated fatty acids on T cell activation and in the modulation of immune system functions by certain dietary lipids. Among these proteins, intracellular Src family protein-tyrosine kinases, transmembrane linker for activation of T cells or interferon-receptors are included in these lipid microdomains and affect T cell signalling [40,62,79].

Overall, several lines of evidence have determined that long-chain *n*-3 polyunsaturated fatty acids, such as eicosapentaenoic acid, are incorporated into cell membranes, replacing arachidonic acid (the most important fatty acid of the eicosanoid precursors). This process reduces the production of an important biological mediator as prostaglandin E_2 responsible for inhibiting interleukin-1 (IL-1) and tumor necrosis factor (TNF) production [20]. By a strict contrast, eicosapentaenoic acid gives rise to 3-series prostaglandins and 5-series leukotrienes, which are lesser reactive than those derived from arachidonic acid. Similarly, eicosapentaenoic acid inhibits gene transcription because it reduces the translocation of nuclear factor κB (NF-κB) [41]. Therefore, fatty acids may exert a direct or an indirect effect on the cells that compose the immune system in both animals and humans. Thus, dietary lipid administration appears to affect the above exposed mechanisms, which will be discussed in more detail.

3.2.1. Alteration of Membrane Fluidity

Changes in the fatty acid composition modify cell membrane fluidity by the administration of certain dietary lipids [65]. The administration of unsaturated fatty acid diets increases the fluidity of plasma membrane, whereas saturated fatty acid diets reduce the fluidity of cell membrane. The alteration of phospholipids composition from plasma membrane represents an important factor responsible for the modification of crucial proteins associated with this cellular structure. As a direct result, numerous cellular functions may be affected, such as intercellular interaction, receptor expression or signal transduction. Thus, the bind of numerous cytokines to their respective receptor placed in the membrane surface may be altered [23]. Similarly, the expression of major histocompatibility complex class II molecules and intercellular adhesion molecule- 1 (ICAM-1) is modified after dietary lipid administration, which lead to a significant inhibition of antigen-presenting function. This fact may be responsible, in part, for the suppression of cell-mediated immune response associated with the administration of *n*-3 polyunsaturated fatty acids [26,27]. In addition, alteration of membrane fluidity appears to be related to the metastatic capacity of tumor cells [68], which may explain the contribution of certain dietary lipids in the promotion/reduction of cancer and their use as chemoprotective agents.

3.2.2. Eicosanoid Synthesis

Eicosanoids comprise an important family of oxygenated derivatives of dihomo-γ-linolenic, arachidonic or eicosapentaenoic acids that act as a second group of chemical messengers within the immune system. The most important eicosanoids are called prostanoids (prostaglandins [PGs] and tromboxanes [Txs]), leukotrienes (LTs), hydroperoxy-eicosatetraenoic acids (HPETEs), hydroxy-eicosatetraenoic acids (HETEs) and lipoxins. Two enzymes acquire a great importance in the biosynthesis of eicosanoids: cyclooxigenase, which produces the PGs and Txs and lipooxigenase, which produces the HPETEs, HETEs and lipoxins. In general, eicosanoids exert numerous functions and particularly the most important is the participation of these lipid mediators in the regulation of inflammatory processes and therefore are also associated with immunomodulatory effects [21]. As mentioned above, fatty acid precursors are released of plasma membrane and undergo enzymatic degradation to produce eicosanoids, which participate as important lipid mediators. Eicosanoids are capable of regulating pro-inflammatory processes, such as induction of fever, vasodilatation, macrophage- or lymphocyte-derived cytokine production, or differentiation of both T and B lymphocytes [8,21,29,59]. Overall, eicosanoids are involved in the modulation of inflammatory responses and in the regulation of immunity [9,38]. The administration of diets containing fish oil (which are mainly composed by eicosapentaenoic or docosahexaenoic acids) to both humans or animals produces a significant reduction in the amount of arachidonic acid in the phospholipid membrane of cells that participate in the immune functions. This event modifies the production of different eicosanoid types and, therefore the immune system functions are also altered. Therefore, depending on the type of eicosanoids, different functions upon immune system have been associated to be modified. In general, eicosanoids derived from *n*-3 polyunsaturated fatty acids are less reactive than eicosanoids derived from *n*-6 polyunsaturated fatty acids. Thus, prostaglandin E_3 and leukotriene B_5 (both derived from eicosapentaenoic acid) are more potent than eicosanoids derived from arachidonic acid (prostaglandin E_2 and leukotriene B_4) in inhibiting lymphocyte proliferation [63]. In addition, prostaglandins inhibit production of interleukin-1 and tumor necrosis factor [42], whereas leukotrienes increase NK cell activity, interleukin-1 production by macrophages and interferon-γ production by lymphocytes [39,58]. Therefore, fatty acids derived from fish oil may inhibit the metabolism of arachidonic acid by two pathways:

i. reduction of substrate availability, because the administration of *n*-3 polyunsaturated fatty acids results in a significant decrease in the amount of arachidonic acid in plasma membrane of immune system cells.
ii. competence with arachidonic acid to serve as a substrate for both cyclooxygenase and lipooxygenase.

3.2.3. Oxidative Alteration

Lipid peroxidation modifies the expression of surface molecules [65]. This fact may be related to the suppression of HLA-DR expression due to free radical production [24]. The antioxidant nutrient status of the subjects is of critical relevance in the determination of the effects of polyunsaturated fatty acids on immune system functions, because of the incorporation of polyunsaturated fatty acids into plasma membrane increases lipid peroxidation and the requirements for antioxidants substances that protects plasma membrane from lipid peroxidation.

Indeed, an interesting report has suggested that the immunosuppression attributed to certain dietary lipids might be in part associated with an increase of lipid peroxidation, and consequently a substantial reduction of antioxidants (particularly vitamin E) [72]. Therefore, the reported suppressive effects of long-chain *n*-3 polyunsaturated fatty acid administration on immune functions can be prevented by vitamin E supplementation [72], but other antioxidants as N-acetyl-L-cysteine has not shown any protective effect [52].

3.2.4. Regulation of Gene Expression

Evidence from recent studies has indicated the participation of certain fatty acids in the activation of one or more transcription factors through direct action on the intracellular signalling pathways. Thus, recent studies have indicated that the modulatory properties of *n*-3 polyunsaturated fatty acids are due, in part, to a reduction in the activity of nuclear factor κB (NF-κB), which is an important transcription factor involved in the promotion of several pro-inflammatory cytokines, cyclooxygenase and adhesion molecules. NF-κB is an important protein composed by different subunits. Thus, a reduction of NF-κB levels has been documented in LPS-stimulated spleen lymphocytes from mice fed a diet containing fish oil [73]. The exact mechanism of action responsible for these effects is not exactly known, but a decrease in the phosphorilation of inhibitory subunit of NF-κB (IκB) appears to be considered the most probable argument responsible for the inhibition in the levels of this factor [10]. Others transcription factors including peroxisome proliferator-activated receptors activators (PPAR) have been proposed to participate in the expression of inflammatory genes [28], as well as in the modulation of different types of cytokines such as IL-2 and IFN-γ [13]. It is probable that certain fatty acids also regulate the activation of these proteins.

3.2.5. Apoptosis

Programmed cell death or apoptosis has been defined in the last years as an essential mechanism responsible for the regulation of homeostasis, tissue development or immune functions. Thus, damaged, aberrant or unnecessary cells must be eliminated to ensure the correct development of multicellular organisms. Irrespective of its crucial role in normal cell control, apoptosis regulates pathological processes including human clinical disorders such as cancer, autoimmune diseases, viral or bacterial infections and neurodegenerative disorders. Several biological (many viruses, bacteria or parasites), chemical (glucocorticoids, ceramides, etc.) or physical factors (irradiation) are responsible for apoptosis induction [50]. Moreover, different studies performed by *in vitro* assays or by administration of dietary lipids in both animals and humans have recently described an important role of several fatty acids in the induction or inhibition of apoptosis. As a consequence, different mechanisms of action have been proposed in order to explain the action of fatty acids on apoptosis modulation. Thus, polyunsaturated fatty acids such as eicosapentaenoic, docosahexaenoic, saturated fatty acids such as palmitic acid, or fats such as fish oil administered in the diets have been defined as substances capable of inducing cell death via a mitochondrial process [17], or by downregulation of bcl-2 [11,57]. Different mechanisms have been proposed in order to explain the induction of apoptosis by certain fatty acids. Thus, palmitic acid (a saturated fatty acid) added to cellular cultures in *in vitro* assays induces apoptosis via a direct effect on mitochondria because it causes a dissipation of the mitochondrial transmembrane potential ($\Delta\Psi_m$), an event that precedes nuclear apoptosis [17]. In addition, another possibility that explains in part the modulation of apoptosis by long-chain *n*-3 polyunsaturated fatty acids is caused by direct action of these substances on the cells and by activation of the caspase cascade through cytochrome *c* release coupled with a modulation of mitochondrial membrane depolarization [2]. Recent studies have also determined the crucial importance of dietary fatty acids in the reduction of Bcl-2 expression as well as an increase of Fas ligand (Fas-L) expression. Therefore, when the concentration of polyunsaturated fatty acids augments, Bcl-2 expression is reduced and cell death occurs [57,69]. In other words, long-chain *n*-3 polyunsaturated fatty acids contained in fish oil trigger apoptosis by suppressing Bcl-2 expression and increasing Fas-L expression. On the other hand, it has been recently determined that Ras membrane localization (a protein that plays a critical role in the cell growth and apoptosis) is reduced after fish oil administration, which indicates that dietary fish oil may play a protective role against colon cancer development [12]. Recently, a novel mechanism whereby fatty acids, particularly *n*-3 polyunsaturated fatty acids, has been identified to establish a relationship between fatty acids and apoptosis. This mechanism suggests that *n*-3 fatty acids modulate T-cell-mediated immunity by selective deletion of Th1-like cells, while maintaining or increasing the Th2-mediated humoral immune response [67]. Finally, it is important to note that lipid peroxidation exerts a crucial importance in apoptosis induction. This preliminary argument indicates that the effects of polyunsaturated fatty acids may be directly attributed to specific alterations on the expression of Bcl-2, Fas and Ras [14].

3.2.6. Other Hypothetical Considerations

In addition of these biochemical events related to the action of fatty acids on the immune functions, a crucial role of lymphocyte subsets has been suggested to explain the effects of different dietary polyunsaturated fatty acids upon the immune system, because they modify the number of lymphocyte subsets as well as the proliferation of these cells [33,71]. Thus, antigen presentation and proportion of T cell subsets are altered after dietary lipid manipulation. Indeed, recent studies have determined that polyunsaturated fatty acids inhibit the surface expression of major histocompatibility complex (MHC) class II molecules, as well as some adhesion molecules on human monocytes [26,27]. Similarly, consumption of a monounsaturated fatty acid-rich diet by humans is also involved in the reduction of adhesion molecules from peripheral blood mononuclear cells [77]. Overall, these actions may explain the different expression of CD4 or CD8 on the T cell surface from peripheral blood of mice that were fed diets supplemented with docosahexaenoic acid [61]. The biological consequences that appear as a result of these changes are still unclear, but it is probable that the alteration experimented by lymphocyte population enhance the host susceptibility against opportunistic infections. In other words, the modulation of lymphocyte subsets may lead in part to a modification of cytokine production, NK cell activity, antibody production, etc. Accordingly, Th1-type cytokines such as IL-2, are more susceptible to polyunsaturated fatty acid effects than are Th2-type cytokines such as IL-4, which indicates that these fatty acids are capable of altering the Th1-/Th2 type cytokine balance in mice [70]. As a direct consequence derived from a reduction of IL-2 production, Th1 cell activation is suppressed, whereas an increase of Th2 cell activation due to an enhancement of IL-4 production suppresses Th1 cell proliferation in animals fed a diet containing fish oil [3]. Therefore, the anti-inflammatory effects derived from long-chain *n*-3 polyunsaturated fatty acid administration may be the consequence of these two interconnected mechanisms. Hence, it is obvious that these alterations may affect significantly the interaction between host natural resistance and infectious agents due to the crucial importance of Th1-mediated response in the infectious processes promoted by viruses or bacteria [47]. Although results from epidemiological, experimental or clinical studies have widely demonstrated an important association between *n*-3 polyunsaturated fatty acids and immunosuppressive properties in both humans and animals, their effects are complex and still not clearly known. In fact, different discrepant results have suggested that *n*-3 polyunsaturated fatty acids have stimulatory effects on TNF and IL-1 production [49].

3.3. Modulation of Immune System Functions by Dietary Lipids

3.3.1. Lymphocyte Proliferation

The alteration of several immune functions associated with fatty acid administration includes the reduction of lymphocyte proliferation, which is mainly modified by polyunsaturated fatty acids (*n*-3 or *n*-6) or monounsaturated fatty acids (*n*-9). Studies carried out in both humans and animals have revealed that the administration of high levels of dietary *n*-3 polyunsaturated fatty acids or the inclusion of lipid emulsions in parenteral regimens rich in *n*-3 or *n*-9 fatty acids are related to the reduction of lymphocyte proliferation during the supplementation [20,22,44,48,74]. Thus, concanavalin-A (Con A) or lipopolysaccharide (LPS)-stimulated lymphocytes have reduced the cellular proliferation in assays carried out *in vitro* or *ex vivo*, in the presence of free fatty acids or in cell cultures from both animals and humans fed dietary lipids, respectively [6,20,56]. Generally, it should be noted that diets containing long-chain *n*-3 polyunsaturated fatty acids, such as docosahexaenoic or eicosapentaenoic acids, suppress the mitogenic response of lymphocytes to a greater extent than diets rich in saturated fatty acids [6]. The effect of fish oil diets on lymphocyte proliferation affects lymphocyte subpopulations. A comparative study in rats fed diets containing fish oil and other types of fatty acids showed that a significant difference was observed between lymphocyte suppression in spleen from rats and thymus, in which lymphocyte proliferation is much lower [74]. Most studies investigating the effects of fatty acids on the immune system have demonstrated that a diet containing olive oil administered to rats promotes a significant reduction of lymphocyte proliferation in response to the mitogen concanavalin A (Con A) [74]. At a first glance, olive oil is composed by different types of substances, and therefore anyone of them may be associated with the suppression of lymphocyte proliferation. On the basis of this hypothesis, an interesting study examined the biological component of olive oil responsible for this suppression, because olive oil contains sterols, alcohols and antioxidants apart from oleic acid and other less relevant fatty acids. This investigation revealed that the immunosuppressive effects of olive oil are likely due to oleic acid rather than to other components of this fat [32]. However, this aspect is not fully demonstrated. By contrast, human studies have produced conflicting results, because a diet rich in olive oil does not affect proliferation of mitogen-stimulated leukocytes [77]. It is possible that the differences between animal and human response may be attributed to the amount of monounsaturated fat employed in the animal studies, where diets contributed approximately 30 % of total energy, whereas in human studies diets supplied approximately 18 % of total energy intake [75]. Although the mentioned studies demonstrated a significant reduction of lymphocyte proliferation in animals, we did not find significant differences in the proliferation of lymphocytes from mice fed with a diet containing olive oil and stimulated with Con A (which induces proliferation of T lymphocytes) or lipopolysaccharide (LPS) (which induces proliferation of B lymphocytes) [55].

It is very interesting to remark on the biological consequences in the clinical application of certain fatty acids on the lymphocyte proliferation. These processes appear to play a crucial role in the amelioration of patients suffering from autoimmune diseases or inflammatory disorders (rheumatoid arthritis, Crohn´s disease or systemic lupus erythematosus), who are characterized by an overactivation of immune system [25].

3.3.2. Cytokine Secretion

Cytokine production is commonly modulated by the administration of certain dietary lipids and particularly long-chain *n*-3 polyunsaturated fatty acids. In general, cytokines may be divided into lymphokines and monokines. The first one are involved in specific immunity and produced by T cells in an activation state, whereas the second one are synthesized by macrophages after stimulation. These proteins regulate the growth and differentiation of different lymphocyte subsets and they activate and regulate cells that participate in the inflammatory response. The reduction of lymphocyte proliferation related to fatty acid administration is associated with the decrease of cytokine production detected after *n*-3 polyunsaturated fatty acid diet administration [76]. Several studies have demonstrated that the immunomodulatory properties of polyunsaturated or monounsaturated fatty acids affect the reduction of pro- and anti-inflammatory cytokines. The most immunosuppressant fatty acids are derived from fish oil and they have been demonstrated to be equipotent in their capacity to reduce IL-2 production [34]. Indeed, the enhancement of fish oil in the diet of humans leads to a significant inhibition of IL-2 production [44]. In addition to IL-2, the production of other cytokines has been investigated. Thus, the secretion of pro-inflammatory cytokines such as IL-1 or TNF are reduced after fish oil diet administration in humans [20], whereas another study has demonstrated that dietary fish oil increases TNF production, however, reduces the secretion of IL-10 (an anti-inflammatory cytokine) in murine peritoneal macrophages [51]. Taking into account the relevant role of T cells in appropriate immune responses to pathogenic agents and autoimmune diseases is important to consider the Th1-type or Th2-type cytokine production after dietary lipid administration. In this context, a recent and interesting investigation indicates that *n*-3 polyunsaturated fatty acids exert a profound inhibition of Th1-type cytokines, whereas they have little effects on Th2-type cytokines [70]. Other dietary fatty acids such as those included in olive oil diet are also involved in the modulation of cytokine production. In fact, the administration of diets containing olive oil for 8 or 12 weeks produced an increase of IL-2 production [16,76] and the synthesis of IL-12, a cytokine that participates as a pro-inflammatory protein was reduced in mice fed with diets containing olive oil [54]. In spite of that olive oil diet is involved in the reduction of cytokine secretion, their immunosuppressant effects are not as potent as those exerted by the administration of a fish oil diet [53].

3.3.3. Natural Killer Cell Activity

The most important function of immune system is the recognition, degradation and elimination of foreign agents. However, target cells also include transformed cells or cells invaded by intracellular bacteria or viruses. These last events are regulated by a type of cells as natural killer (NK) cells that are a subset of lymphocytes found in blood and spleen, derived from bone marrow, but neither T cells nor B cells. In fact, it has been demonstrated that a reduction in the total fatty acids consumption in the diet produces an increase of NK cell activity [4]. In animal studies, olive oil diet administration reduced the NK cell activity, although the most suppressive diet associated with an important decrease of NK cell activity from rats was that containing fish oil [28,31]. In fact, NK cell activity was significantly reduced in mice fed with olive oil after receiving tumor transplantation LSTRA, although the greatest suppression was again observed in mice fed with a fish oil diet, which appears to be involved in a relevant reduction of mice survival after tumor transplantation, because all of the mice fed with this diet died at 5th day of tumor injection [55]. The comparison of different doses of oleic acid in the diet showed a negative linear relationship, indicating that oleic acid is responsible for a reduction of NK cell activity in rats [31]. Once again, human studies have not confirmed similar results, because the measurement of NK cell activity after the administration for one or two months of a diet containing olive oil did not show relevant differences, although the activity of these cells declined after two months of the incorporation of the mentioned diet [77].

3.3.4. Adhesion Molecules

An interesting line of investigation is based on the potential effect of dietary lipids on the expression of adhesion molecules, which represent an important factor in the basic mechanisms of inflammation and in the cell-cell interactions. Thus, several studies have demonstrated that there was a decrease in the levels of expression of CD-2, lymphocyte function antigen-1 (LFA-1) and intercellular adhesion molecule-1 (ICAM-1, CD54) on spleen lymphocytes from rats fed with diets containing olive oil or fish oil [60]. Similarly, an olive oil diet reduced the levels of ICAM-1 after two months of dietary administration in humans [74,76]. Other studies showed that culture of human venous endothelial cells in the presence of eicosapentaenoic or docosahexaenoic acids reduced LPS- or cytokine-induced surface expression of E-selectin, ICAM-1 and vascular cell adhesion molecule-1 (VCAM-1) [15,37]. Indeed, the downregulation of VCAM-1 expression carried out by docosahexaenoic acid was exerted in an independent form of eicosanoid production and antioxidant status, but directly at the level of VCAM-1 gene expression [15]. Dietary fish oil also reduces the expression of ICAM-1 on the surface of murine macrophages [46], or circulating levels of soluble VCAM-1 in humans [45]. These relevant studies have an important clinical interpretation, because ICAM-1 is an adhesion molecule involved in different functions. This molecule participates in leukocyte-leukocyte adhesion as well as in leukocyte-endothelial cells adhesion. This molecule is also expressed on mononuclear cells isolated from inflamed synovium in individuals suffering from rheumatoid arthritis. In addition, it also participates in the recruitment of mononuclear cells responsible for the formation and development of atherosclerotic plaque.

4. Conclusion

Despite extensive research during the last thirty years about the effects promoted on immune system of both humans and animals by dietary lipids, recent advances have revealed numerous discrepancies supported by experimental and clinical studies, which have confirmed that the action of dietary lipid manipulation on immune system functions depends on different factors, such as type of cells, class of fatty acid, concentration, methodological studies, etc. Therefore, in the light of current experimental observations, it is clear that certain fatty acids modulate the immune system in both animals and humans, and various mechanisms of action have been proposed to explain the action of fatty acids on different functions of immune cells. Thus, it is generally accepted that lymphocyte proliferation, cytokine synthesis, NK cell activity or adhesion molecules are immune parameters affected after the administration of diets containing *n*-3 polyunsaturated fatty acids or *n*-9 monounsaturated fatty acids, whereas *n*-6 polyunsaturated fatty acids does not exert a suppression of immune response, at least as efficient as *n*-3 polyunsaturated fatty acids. As a direct consequence of these events, certain fatty acids have been applied in the reduction of inflammatory disorders, which are characteristics of autoimmune diseases, such as rheumatoid arthritis, inflammatory bowel disease, psoriasis or systemic lupus erythematosus. In addition, other epidemiological and clinical studies have indicated the intervention of these dietary lipids as chemopreventive and chemoprotective substances in breast, colorectal and prostate cancers. Nevertheless, the immunosuppression carried out by certain unsaturated fatty acids (particularly long-chain *n*-3 polyunsaturated fatty acids, the most immunosuppressive fatty acids) may lead to a significant increase of susceptibility to infectious diseases, or in other words, to a loss of host immune resistance against viruses, bacteria, fungi or parasites. On the contrary, other studies have suggested that long-chain *n*-3 polyunsaturated fatty acids improve the immune resistance, and therefore significantly reduce the susceptibility of both animals and humans against infectious microorganisms. Hence, these discrepancies should be discussed in a further chapter. Finally, when investigating the action of dietary lipids or fatty acids on immune system and inflammatory processes, the interaction of fatty acids with other nutrients (for example vitamin E) must be taken into account. Therefore, it is clear that the future investigation should be focused on the relationship between several fatty acids and the potential consequences in the clinical application of these nutrients, because several parenteral lipid emulsions are currently applied in clinical nutrition as a potential alternative to others classically used.

References

[1] Allard, JP; Kurian, R; Aghdassi, E; Muggli, R; Royall, D. Lipid peroxidation during *n*-3 fatty acid and vitamin E supplementation in humans. *Lipids,* 1997 32, 535-541.

[2] Arita, K; Kobuchic, H; Utsumia, T; Takeharab, Y; Akiyamad, J; Hortone, AA; Utsumib, K. Mechanism of apoptosis in HL-60 cells induced by *n*-3 and *n*-6 polyunsaturated fatty acids. *Biochemical Pharmacology,* 2001 62, 821–828.

[3] Arrington, JL; Chapkin, RS; Switzer, KC; Morris, JS; McMurray, DN. Dietary *n*-3 polyunsaturated fatty acids modulate purified murine T-cell subset activation. *Clinical and Experimental Immunology,* 2001 125, 499-507.

[4] Barone, J; Helbert, JR; Reddy, MM. Dietary fat and natural killer cell activity. *American Journal of Clinical Nutrition*, 1989 50, 861-867.

[5] Brown, DA; London, E. Functions of lipid rafts in biological membranes. *Annual Review of Cell and Developmental Biology*, 1998 14, 111-136.

[6] Calder, PC. Fatty acids, dietary lipids and lymphocyte functions. *Biochemical Society Transaction*, 1995 23, 302-309.

[7] Calder, PC. Fat chance of immunomodulation. *Immunology Today*, 1998 19, 244-247.

[8] Calder, PC. Polyunsaturated fatty acids, inflammation and immunity. *Lipids,* 2001 36, 1007-1024.

[9] Calder, PC. *N*-3 polyunsaturated fatty acids and inflammation: from molecular biology to the clinic. *Lipids*, 2003 38, 343-352.

[10] Chen, LH; Zhao, Y. Eicosapentaenoic acid decreases lipopolysaccharide-stimulated tumor necrosis factor-α expression by inhibiting nuclear factor kB activation. *FASEB Journal*, 2001 15: A258.

[11] Chiu, LC; Wan, JM. Induction of apoptosis in HL-60 cells by eicosapentaenoic acid (EPA) is associated with downregulation of bcl-2 expression. *Cancer Letters*, 1999 145, 17-27.

[12] Collett, ED, Davidson, LA; Fan, YY; Lupton, JR; Chapkin, RS. *N*-6 and *n*-3 polyunsaturated fatty acids differentially modulate oncogenic Ras activation in colonocytes. *American Journal of Physiology-Cell Physiology*, 2001 280, 1066-1075.

[13] Cunard, R; Ricote, M; DiCampli, D; Archer, DC; Kahn, DA; Glass, CK; Kelly, CJ. Regulation of cytokine expression by ligands of peroxisome proliferator activated receptors. *The Journal of Immunology*, 2002 168, 2795-2802.

[14] Das, UN. Essential fatty acids, lipid peroxidation and apoptosis. *Prostaglandins, Leukotrienes and Essentials. Fatty Acids*, 1999 61; 157-163.

[15] De Caterina, R; Cybulsky, ML; Clinton, SK; Gimbrone, MA; Libby, P. The omega-3 fatty acid docosahexaenoate reduces cytokine-induced expression of protherogenic and proinflammatory proteins in human endothelial cells. *Arteriosclerosis and Thrombosis*, 1994 14, 1829-1836.

[16] de Pablo, MA; Ortega, E; Gallego, AM; Alvarez, C; Pancorbo, PL; Alvarez de Cienfuegos, G. Influence of diets containing olive oil, sunflower oil or hydrogenated coconut oil on the immune response of mice. *Journal of Clinical Biochemical Nutrition*, 1998 25, 11-23.

[17] de Pablo, MA; Susin, SA; Jacotot, E; Larochette, N; Costantini, P; Ravagnan, L; Zanzami, N; Kroemer, G. Palmitate induces apoptosis via a direct effect on mitochondria. *Apoptosis*, 1999 4, 81-87.

[18] de Pablo, MA; Alvarez de Cienfuegos, G. Modulatory effects of dietary lipids on immune system functions. *Immunology and Cell Biology*, 2000 78, 31-39.

[19] de Pablo, MA; Puertollano, MA; Alvarez de Cienfuegos, G. Biological and clinical significance of lipids as modulators of immune system functions. *Clinical and Diagnostic Laboratory Immunology*, 2002 9, 945-950.

[20] Endres, S; Ghorbani, R; Kelley, VE; Georgilis, K; Lonemann, G; van der Meer, JMW; Cannon, JG; Rogers, TS; Klempner, MS; Weber, PC; Schaeffer, EJ; Wolff, MS; Dinarello, CA. The effect of dietary supplementation with *n*-3 polyunsaturated fatty acids on the synthesis of interleukin-1 and tumor necrosis factor by mononuclear cells. *New England Journal of Medicine*, 1989 320, 265-271

[21] Goodwin, JS; Ceuppens, J. The regulation of immune responses by prostaglandins. *Journal of Clinical Immunology*, 1983 3, 295-308.

[22] Granato, D; Blum, S; Rössle, C; Le Boucher, J; Malnoë, A; Dutot, G. Effects of parenteral lipid emulsions with different fatty acid composition on immune cell functions *in vitro*. *Journal Parenteral and Enteral Nutrition*, 2000 24, 113-118.

[23] Grimble, RF; Tappia, PS. The modulatory influence of unsaturated fatty acid on the biology of tumor necrosis factor. *Biochemical Society Transaction*, 1995 287, 282-286.

[24] Gruner, S; Volk, HD; Falck, P; Baehr, RV. The influence of phagocytic stimuli on the expression of HLA-DR antigens: role of reactive oxygen intermediates. *European. Journal of Immunology*, 1986 16, 212-215.

[25] Harbige, LS. Fatty acids, the immune response, and autoimmunity: a question of *n*-6 essentiality and the balance between *n*-6 and *n*-3. *Lipids*, 2003 38, 323-341.

[26] Hughes, DA; Pinder, AC; Piper, Z; Johnson, IT; Lund, EK. Fish oil supplementation inhibits the expression of major histocompatibility complex class II molecules and adhesion molecules on human monocytes. *American Journal of Clinical Nutrition*, 1996 63, 267-272.

[27] Hughes, DA; Pinder, AC. *N*-3 polyunsaturated fatty acids inhibit the antigen-presenting function of human monocytes. *American Journal of Clinical Nutrition*, 2000 71, 357S-360S.

[28] Hwang, D; Rhee, SH. Receptor-mediated signalling pathways: potential targets of modulation by dietary fatty acids. *American Journal of Clinical Nutrition*, 1999 70, 545-556.

[29] Hwang, D. Essential fatty acids and immune response. *FASEB Journal*, 1989 3, 2052-2061.

[30] Janeway, CA. How the immune system protects the host from infection. *Microbes and Infection*, 2001 3, 1167-1171.

[31] Jeffery, NM; Cortina, M; Newsholme, EA; Calder, PC. Effects of variations in the proportions of saturated, monounsaturated and polyunsaturated fatty acids in the rat diet on spleen lymphocyte functions. *British Journal of Nutrition,* 1997 77, 805-823.

[32] Jeffery, NM; Yaqoob, P; Newsholme, EA; Calder, PC. The effects of olive oil upon rat serum lipid levels and lymphocyte functions are due to oleic acid. *Annals of Nutrition and Metabolism*, 1997 40, 71-80.

[33] Jenski, LJ; Bowker, GM; Johnson, MA; Ehringer, WD; Fetterhoff, T; Stillwell, W. Docosahexaenoic acid-induced alteration of Thy-1 and CD8 expression on murine splenocytes. *Biochimica et Biophysica Acta*, 1995 1236, 39-50.

[34] Jolly, CA; Jiang, YH; Chapkin, RS; McMurray, DN. Dietary *n*-3 polyunsaturated fatty acids suppress murine lymphoproliferation, interleukin-2 secretion and the formation of diacylglycerol and ceramide. *The Journal of Nutrition*, 1997 127, 37-43.

[35] Jump, DB; Clarke, SD. Regulation of gene expression by dietary fat. *Annuals Review of Nutrition*, 1999 19, 63-90.

[36] Kelley, DS. Modulation of human immune and inflammatory responses by dietary fatty acids. *Nutrition*, 2001 17, 669-673.

[37] Kim, DN; Schmee, J; Thomas, WA. Dietary fish oil added to a hyperlipidemic diet for swine results in reduction in the excessive number of monocytes attached to the arterial endothelium. *Atherosclerosis,* 1995 81, 209-216.

[38] Kinsella, JE; Lokesh, B; Broughton, S; Whelan, J. Dietary polyunsaturated fatty acids and eicosanoids: potential effects on the modulation of inflammatory and immune cells: an overview. *Nutrition*, 1990 6, 24-44.

[39] Lewis, RA. Interactions of eicosanoids and cytokines in immune regulation. *Advances of Prostaglandin, Thromboxane and Leukotriene Research,* 1990 20, 170-178

[40] Liang, X; Nazarian, A; Erdjument-Bromage, H; Bornmann, W; Tempst, P; Resh, MD. Heterogeneous fatty acylation of Src family kinases with

polyunsaturated fatty acids regulates raft localization and signal transduction. *Journal of Biological Chemistry*, 2001 276, 30987-30994.

[41] Lo, CJ; Chiu, KC; Fu, M; Lo, R; Helton, S. Fish oil decreases macrophage tumor necrosis factor gene transcription by altering Nfkappa B activity. *Journal of Surgery Research*, 1999 82, 216-221.

[42] Marcinkiewicz, J. *In vitro* cytokine release by activated murine peritoneal macrophages: role of prostaglandins in the differential regulation of tumor necrosis factor alpha, interleukin 1, and interleukin 6. *Cytokine*, 1991 3, 327-332.

[43] McMurray, DN, Jolly, CA; Chapkin, RS. Effects of dietary *n*-3 fatty acids on T cell activation and T cell receptor-mediated signalling in a murine model. *The Journal of Infectious Diseases*, 2000 182, S103-S107.

[44] Meydani, SN; Endres, S; Woods, MM; Goldin, BR; Soo, C; Morrill-Labrode, A; Dinarello, C; Gorbach, SL. Oral *n*-3 fatty acid supplementation suppresses cytokine production and lymphocyte proliferation: comparison between young and older women. *The Journal of Nutrition*, 1991 121, 547-555.

[45] Miles, EA; Thies, F; Wallace, FA; Powell, JR; Hirst, TL; Newsholme, EA; Calder, PC. Influence of age and dietary fish oil on plasma soluble adhesion molecule concentration. *Clinical Science*, 2001 100, 91-100.

[46] Miles, EA; Wallace, FA; Calder, PC. Dietary fish oil reduces intercellular adhesion molecule 1 and scavenger receptor expression on murine macrophages. *Atherosclerosis*, 2000 152, 43-50.

[47] Mosmann, TR; Sad, S. The expanding universe of T-cell subsets: Th1, Th2 and more. *Immunology Today*, 1996 17, 138-146.

[48] Moussa, M; Le Boucher, J; Garcia, J; Tkaczuk, J; Ragab, J; Dutot, G; Ohayon, E; Ghisolfi, J; Thouvenot, JP. *In vivo* effects of olive oil-based lipid emulsion on lymphocyte activation in rats. *Clinical Nutrition* 2000 19, 49-54.

[49] Netea, MG; Kullberg, BJ; Blok, WL; Van der Meer, JWM. Immunomodulation by *n*-3 polyunsaturated fatty acids. *Immunology Today*, 1999 20, 103.

[50] Penninger, JM; Kroemer, G. Molecular and cellular mechanisms of T lymphocyte apoptosis. *Advances in Immunology*, 1998 68, 51-144.

[51] Petursdottir, DH; Olafsdottir, I, Hardardottir, I. Dietary fish oil increases tumor necrosis factor secretion but decreases interleukin-10 secretion by murine peritoneal macrophages. *The Journal of Nutrition*, 2002 132, 3740-3743.

[52] Puertollano, MA; de Pablo, MA; Álvarez de Cienfuegos, G. Anti-oxidant properties of N-acetyl-L-cysteine do not improve the immune resistance of mice fed dietary lipids to *Listeria monocytogenes* infection. *Clinical Nutrition*, 2003 22, 313-319.
[53] Puertollano, MA; de Pablo, MA; Alvarez de Cienfuegos, G. Olive oil and immune system functions: potential involvement in immunonutrition. *Grasas y Aceites*, 2004 55, 42-51.
[54] Puertollano, MA; Puertollano, E; Ruiz-Bravo, A; Jiménez- Valera, M; de Pablo, MA; Alvarez de Cienfuegos, G. Changes in the immune functions and susceptibility to *Listeria monocytogenes* infection in mice fed dietary lipids. *Immunology and Cell Biology*, 2004 82, 370-376.
[55] Puertollano, MA; Algarra, I; Ortega, E; de Pablo, MA; Alvarez de Cienfuegos, G. Loss of cell natural killer cell activity after murine tumor transplantation appears as a consequence of dietary lipid administration. *Anticancer Research*, 2001 21, 2697-2702.
[56] Puertollano, MA; de Pablo, MA; Alvarez de Cienfuegos, G. Relevance of dietary lipids as modulators of immune functions in cells infected with Listeria monocytogenes. *Clinical and Diagnostic Laboratory Immunology*, 2002 9, 352-357.
[57] Reddy Avula, CP; Zaman, AK; Lawrence, R; Fernandes, G. Induction of apoptotic mediators in Balb/c splenic lymphocytes by dietary *n*-3 and *n*-6 fatty acids. *Lipids*, 1999 34, 921-927.
[58] Rola-Pleszczynski, M; Galgnon, L; Sirois, P. Leukotriene B4 augments human natural cytotoxic cell activity. *Biochemical and Biophysical Research Communications*, 1983 113, 531-537.
[59] Rola-Pleszczynski, M; Stankova, J. Cytokine gene regulation by PGE2, LTB4 and PAF. *Mediators of Inflammation*, 1992 1, 5-8.
[60] Sanderson, P; Yaqoob, P; Calder, PC. Effects of dietary lipid manipulation upon graft vs host and host vs graft responses in the rat. *Cellular Immunology*, 1995 164, 240-247.
[61] Sasaki, T; Kanke, Y; Nagahashi, M; Toyokawa, M; Matsuda, M; Shimizu, J; Misawa, Y; Takita, T. Dietary docosahexaenoic acid can alter the surface expression of CD4 and CD8 on T cells in peripheral blood. *Journal of Agricultural and Food Chemistry*, 2000 48, 1047-1049.
[62] Sehgal, PB; Guo, GG; Shah, M; Kumar, V; Patek, K. Cytokine signalling: STATS in plasma membrane rafts. *Journal of Biological Chemistry*, 2002 277, 12067-12074.

[63] Shapiro, AC; Wu, D; Meydani, SN. Eicosanoid derived from arachidonic and eicosapentaenoic acids inhibit T cell proliferative response. *Prostaglandins*, 1993 45, 229-240.

[64] Simons, K; Toomre, D. Lipid rafts and signal transduction. *Nature Reviews. Molecular Cell Biology*, 2000 1, 31-39.

[65] Stubbs, CD; Smith, AD. The modification of mammalian polyunsaturated fatty acid composition in relation to membrane fluidity and function. *Biochimica et Biophysica Acta*, 1984 779, 89-137.

[66] Stulnig, TM; Huber, J; Leitinger, N; Imrel, EM; Angelisova, A; Nowotny, P; Waldhausi, W. Polyunsaturated eicosapentaenoic acid displaces proteins from membrane rafts by altering raft lipid composition. *Journal of Biological Chemistry*, 2001 276, 37335-37340.

[67] Switzer, KC; McMurray, DN; Morris, JS; Chapkin, RS. (*n*-3) polyunsaturated fatty acids promote activation-induced cell death in murine T lymphocytes. *The Journal of Nutrition*, 2003 133, 496-503.

[68] Taraboletti, G; Perin, L; Botazzi, B; Mantovani, A; Giavazzi, R; Salmona, M. Membrane fluidity affects tumor-cell motility, invasion and lung-colonizing potential. *International Journal of Cancer*, 1989 44, 707-713.

[69] Tyurina, YY; Tyurin, VA; Carta, G; Quinn, PJ; Schor, NF; Kagan, VE. Direct evidence for antioxidant effect of Bcl-2 in PC12 rat pheochromocytoma cells. *Archives of Biochemistry and Biophysics*, 1997 344, 413-423.

[70] Wallace, FA; Miles, EA; Evans, C; Stock, TE; Yaqoob, P; Calder, PC. Dietary fatty acids influence the production of Th1- but not Th2-type cytokines. *Journal of Leukocyte Biology*, 2001 69, 449-457.

[71] Wang, YW; Field, CJ; Sim, JS. Dietary polyunsaturated fatty acids alter lymphocyte subset proportion and proliferation, serum immunoglobulin G concentration, and immune tissue development in chicks. *Poultry Sciences*, 2000 79, 1741-1748.

[72] Wu, D; Meydani, SN. *N*-3 polyunsaturated fatty acids and immune function. *The Proceedings of the Nutrition Society*, 1998 57, 503-509.

[73] Xi, S; Cohen, D; Barve, S; Chen, LH. Fish oil suppressed cytokines and nuclear factor kB induced by murine AIDS virus infection. *Nutrition Research*, 2001 21, 865-878.

[74] Yaqoob, P; Newsholme, EA; Calder, PC. The effect of dietary lipid manipulation on rat lymphocyte subsets and proliferation. *Immunology*, 1994 82, 603-610.

[75] Yaqoob, P. Monounsaturated fatty acids and immune function. *European Journal of Clinical Nutrition*, 2002 56, S9-S13.

[76] Yaqoob, P; Calder, PC. The effects of dietary lipid manipulation on the production of murine T cell-derived cytokines. *Cytokine*, 1995 7, 548-553.
[77] Yaqoob, P; Knapper, JA; Webb, DH; Willians, CM; Newsholme, EA; Calder, PC. Effect of olive oil on immune function in middle-aged men. *American Journal of Clinical Nutrition*, 1998 67, 129-135.
[78] Yaqoob, P; Newsholme, EA; Calder, PC. Inhibition of natural killer cell activity by dietary lipids. *Immunology Letters*, 41, 241-247.
[79] Zeyda, M; Staffler, G; Horejsi, V; Waldhausl, W; Stulnig, TM. LAT displacement from lipid rafts as a molecular mechanism for the inhibition of T cell signalling by polyunsaturated fatty acids. *Journal of Biological Chemistry*, 2002 277, 28418-28423.

Chapter Sources

The following chapters have been previously published:

Chapter 5 – A version of this chapter was also published in Lipids: Categories, Biological Functions and Metabolism, Nutrition and Health, edited by Paige L. Gilmore, published by Nova Science Publishers, Inc. It was submitted for appropriate modifications in an effort to encourage wider dissemination of research.

Chapter 6 – A version of this chapter was also published in Trends in Dietary Fats Research, edited by M.V Landow, published by Nova Science Publishers, Inc. It was submitted for appropriate modifications in an effort to encourage wider dissemination of research.

Index

A

D

E

F

G

H

I

K

L

N

O

P

Q

R

S

T

U

V

W

Y